Kauṭilya's Arthaśāstra

— In the Light of Modern Science and Technology —

Reconstructing Indian History & Culture
(ISSN 0971-3824)

1. Society in Ancient India; by Sures Chandra Banerji.
2. Political Thought in Ancient India; by G.P. Singh.
3. Early Indian Historical Tradition and Archaeology; by G.P. Singh.
4. Vasantotsava: The Spring Festivals of India; by Leona M. Anderson.
5. Ecological Readings in the Veda; by M. Vannucci.
6. Vṛndāvana in Vaiṣṇava Literature; by Maura Corcoran.
7. Social History of the Tamils (1707-1947); by P. Subramanian.
8. Ancient Indian Coinage; by Rekha Jain.
9. Buddhism and Socio-Economic Life of Eastern India; by Bimal Chandra Mohapatra.
10. Chandragupta Maurya; by P.L. Bhargava.
11. Mṛtyu: Concept of Death in Indian Traditions; by Gian Giuseppe Filippi.
12. Yama: The Glorious Lord of the Other World; by Kusum P. Merh.
13. Sacred Complex of Ujjain; by D.K. Samanta.
14. Vedic View of the Earth; by S.R.N. Murthy.
15. Society in the Atharvaveda; by B.S. Kharade.
16. Orissan History, Culture and Archaeology; by S. Pradhan.
17. Education in Ancient India; by Mitali Chatterjee.
18. Surā: The Liquor and the Vedic Sacrifice; by M.B. Kolhatkar.
19. Human Ecology in the Vedas; by Marta Vannucci.
20. Governance in Ancient India; by Anupa Chandra Pandey.
21. The Cultural Glory of Ancient India: A Literary Overview; by Sures Chandra Banerji.
22. Kauṭilya's Arthaśāstra in the light of Modern Science & Technology; by Sunil Sen Sarma.
23. History of Jainism — With Special Reference to Mathurā; by V.K. Sharma.
24. Cultural Tourism in India: Museums, Monuments & Arts — Theory and Practice; by S.P. Gupta; Krishna Lal; Mahua Bhattacharya.
25. Ancient Indian Historiography — Sources and Interpretations; by G.P. Singh.
26. Republics, Kingdoms, Towns and Cities in Ancient India; by G.P. Singh.
27. Facets of Ancient Indian History and Culture — New Perception; by G.P. Singh.
28. Kamakhya — A Socio-cultural Study; by Nihar Ranjan Mishra.
29. History and Culture of Tamil Nadu as Gleaned from the Sanskrit Inscriptions Vol. 1 (up to c. AD 1310); by Chithra Madhavan.
30. Lifestyle of the Vedic People; by Pranati Ghosal.
31. History and Culture of Tamil Nadu as Gleaned from the Sanskrit Inscriptions Vol. 2 (up to c. AD 1310 - *c.* 1885 AD); by Chithra Madhavan.
32. Matrimonial Alliances and Ancient Indian Polity (*c.* 600 BCE to *c.* 650); by Preeti Prabhat.
33. Artisans and Craftsmen of Northern India; by Kuldeep Singh Thind.
34. The Evolution of Historiographical Tradition in Ancient and Medieval India; by G.P. Singh.

Reconstructing Indian History & Culture, no. 22

Kauṭilya's Arthaśāstra

— In the Light of Modern Science and Technology —

by

Sunil Sen Sarma

PRINTWORLD

Publishers of Indian Traditions

Cataloging in Publication Data — DK
[Courtesy: D.K. Agencies (P) Ltd. <docinfo@dkagencies.com>]
Sen Sarma, S.B. (Sunil Behari), 1933 —
Kauṭilya's Arthaśāstra.
(Reconstructing Indian history & culture, no. 22).
Includes bibliographical references (p.)
Includes Index
ISBN 13: 9788124601877

1. Kauṭilya. Arthaśāstra. 2. Economics — India — Early works to 1800. 3. Science in literature. 4. Technology in literature. I. Title. II. Series: Reconstructing Indian history & culture, no. 22.

ISBN 13: 978-81-246-0187-7 **ISBN 10: 81-246-0187-9**
First Published in India in 2001
Second Impression in 2012

Published and printed by:
D.K. Printworld (P) Ltd.
Regd. Office: 'Vedasri', F-395, Sudarshan Park
(Metro Station: Ramesh Nagar)
New Delhi-110 015
Phones: (011) 2545 3975; 2546 6019; *Fax*: (011) 2546 5926
E-mail: indology@dkprintworld.com
Website: www.dkprintworld.com

To

My Wife Monika

Acknowledgements

IN presenting my understanding of the science content of *Arthaśāstra*, facts which may be considered as exordium to it relate to the history of my undertaking this work and my indebtedness to personalities who liberally helped me for its completion.

It had a curious and chance beginning. Late Professor Debi Prasad Chattopadhyaya, engaged in his research in the late 1980s on the science of ancient India felt it expedient to incorporate critical analysis of some portions of Kauṭilya's *Arthaśāstra* dealing with gems, minerals, ores, metallurgy, etc., in the light of modern knowledge in those subjects. Knowing my education in geology, I was entrusted with this task. I did not have the opportunity to know, due to his sudden demise, if the short write-up provided to him was of any use. But I was rewarded in being led by him into a new domain hitherto unknown to me, which, I discovered subsequently, is a treasure house of ancient Indian wisdom in diverse subjects. His sudden demise deprived me the opportunity of appraising him the outcome of his enlightening me. I bow in deep reverence to him as the initiator of the present study.

My timorous attempt to work with this archaic literature was emboldened by generous and indulgent encouragement and active help I received from erudite personalities like Late Gopal Haldar and his wife Late Dr. Aruna Haldar. Initial studies were jointly carried out with Dr. Aruna Haldar, who obliged me by agreeing to be the co-author of the first paper on the subject published in the *Vishveshvaranand Indological Journal*, vol. XXVII, parts i-ii, 1989, brought out in 1994. Ever since then, she remained my constant source of inspiration and advisor till her death in June, 1998. I consider myself extremely fortunate as recipient of their illuminating counsel and schooling.

Dr. D.B. Sen Sharma, formerly Chairman, Department of Sanskrit, Pali and Prakrit, Ex-Dean, Faculty of Indic Studies, Kurukshetra University, India, very kindly overviewed the methods of my approach, attended to my probing enquiries on intricacies and significance of language of the *sūtras*, appropriateness and acceptability of their alternative translations, meanings and deconstructions preferred by me, with stoical tolerance of a teacher. But for his help and encouragement, this work could not have been brought to fruition. I received his partronage in various other forms, mainly in projecting this work to a number of people of learning. I owe him my boundless indebtedness.

I am extremely grateful to Dr. Sisir Sen, F.N.A., retired professor of Geology, I.I.T., Kharagpur, who introduced my project to the authorities of the Indian National Science Academy, which enabled me to receive grant for completing the work. I took liberal advantage of his goodness and indulgence towards me throughout my long association with him, for overcoming any problem I faced. I can only express my deep gratitude to him for all these.

To Dr. Asoke Mookherjee, F.N.A., retired professor of Geology I.I.T. Kharagpur, I owe a great deal of thanks for helping me with academic counselling, and obliging me with his ever-ready help to meet any and every odd requirements and demands. So also was Dr. P.K. Banerjee, F.N.A., retired Deputy Director General, Geological Survey of India. I am also obliged to him for his help whenever needed by me.

The list of names who helped me and to whom I am grateful for such helps, is a long one. In spite of that I shall be failing in my duty if I do not mention some of them.

Shri Sibdas Chaudhuri, retired Librarian, The Asiatic Society, Calcutta, showed keen interest in this work, often providing valuable references; Shri Chitrabhanu Sen, retired Librarian of The Asiatic Society, Calcutta, obliged me with his knowledgeable discussions and references on ancient Sankrit literature; Shri S.K. Kapoor, retired Librarian, Geological Survey of India, Calcutta, besides rendering help in the use of library, brought to my notice several publications which were of immense help to me; Shri Jyoti Bhusan Chaki, noted litterateurs, was magnanimous in lending me the *Sanskrit-English Dictionary* by Monier Monier-Wiliams for my constant use. I consider this help as priceless and express my gratitude to him; Dr. Debjani Haldar, Research Scholar, Sanskrit Department, Jadavpur University, allowed me to use her personal collections on Sanskrit literature. I owe them all my sincere thanks.

I acknowledge with thanks the benefits I derived from discussions from time to time with Shri Sudhin Gupta, Managing Director, CASA Engineers International Pvt. Ltd., Calcutta and also with Miss Urmi Sen, Engineer of the same concern, on matters related to civil engineering designs and soil mechanics. I am obliged to Shri Amitava Ghosh, Managing Director, A.G. Consulting Groups, Calcutta, for permitting me to reproduce the engineering interpretation of the design code of towers as contained in *Arthaśāstra*.

Discussions on various occasions on various topics related to the work, with Professor Arun Nag of Viswa Bharati University, Dr. P.K. Chattopadhyaya, an expert on archaeo-metallurgy, of the Alloy Steels Plant, Durgapur, my friend Shri Mihir Sen Gupta, Professor Nrisingha Prasad Bhadhuri enriched and enlightened me. I express my heart-felt thanks to them all.

To the authorities of Indian National Science Academy, particularly to Dr. A.K. Bag, Co-ordinator, History of Science Programmes, I am ever grateful for providing financial grant, enabling me to complete the work.

The authorities of The Asiatic Society, Calcutta obliged me by granting affiliation to the Society to carry out the work under its aegies and providing me the library facilities. I am thankful, particularly to Profesor Amalendu De, Secretary, Dr. G.P. Sandilya, Treasurer, Controller of Accounts and his staff, Dr. Mitali Chatterjee of the Society's Library, Shri Jagatpati Sarkar of the Academic Section, and one and all of the Library staff of The Asiatic Society, who ungrudgingly helped me everytime I required help.

Sarvasri A.J. Baidya, Deputy Superintending Archaeologist, S.K. Ghosal, Assistant Archaeologist, A.K. Bandopadhyaya,

Photographer, Archaeological Survey of India, were kind enough to provide a few photographs of archaeological sites of India. I thank them all.

I am grateful to Professor Raymond Allchin for allowing me to reproduce the map on 'Probable Provincial Groupings of the Mauryan Empire' from his book entitled *The Archaeology of Early Historic South Asia*.

To Late Justice Bijitendra Mohan Mitra, I am indebted for his erudite discussions on various related subjects, particularly on the legal aspects of environment and also for providing me with relevant literature.

Professor Rama Ranjan Mukhopadhyay, formerly Vice-Chancellor of the universities of Bardhaman and Rabindra Bharati, formerly Chairman of the Indian Association of Universities, amongst many other responsible positions in educational institutes and organisations he held and is still holding, obliged me by his generous and inspiring appreciation of the present work. My gratitude to him is boundless.

Dr. Arindam Parui rendered all types of secretarial assistance, including computer composition of manuscript, enduring the troubles and difficulties of effecting changes, modifications, emendations in the manuscript, from the initial stage of drafting to the finalisation of the same. My thanks to him are boundless.

Carefully researched artistic reproduction of fort, royal residence, assembly hall, from literary description of those in the main text and cartographic drawing by Shri Tamal Bhowmick and Sreemati Sanchari Mitra added new dimension to this work. I am indeed thankful to them.

I am extremely fortunate in having the full support of my entire family. My elder daughter, Aparajita Chaudhari looked into the plan and arrangements of the work; Soumya Sen Sarma my son, was like a matured advisor to me in every situation I was confronted with; my youngest daughter, Anamitra Mukherjee, and my niece Swati Sen Sarma were ever available for any kind of help I required; and

Monika, my wife, stood by me with her inherent and endearing poise, tolerance and stimulating encouragements. All of them were almost organically attached with the work.

I must express my gratitude to M/s D.K. Printworld (P) Ltd, particularly its Director, Shri Susheel K. Mittal, for having undertaken the publication.

Calcutta
15 Janaury 2001

Sunil Sen Sarma

Contents

Maps

Photographs and Illustrations

Introduction

THIS work is an attempt to grasp the ancient Indian wisdom of science and technology ingrained in Kauṭiliya's *Arthaśāstra* and ensconced behind its commonly perceived theme of state craft and administration. It is a daring attempt by the present worker with his background as a geologist, having no formal education and grooming in Sanskrit, the language of the source materials. Difficulties were on several counts.

The work started based on various translations of the book. But several confusing and widely different meanings of the same passage or words conveyed in different translations, necessitated frequent journeys into the original text. This could have been due to the fact that translators, mostly Indologists and Scholars in Sanskrit literature, were perhaps more conscious about retaining the literary originality and purity of the original text rather than extracting its esoteric meanings, particularly of scientific and technological topics (Rangarajan, p. 24). Kangle, who did exhaustive translation of the original text and analysed the contents very critically, was also confronted with meanings of a number of words, particularly the technical ones. He feels that

> it is in vocabulary, however, that the language of this work shows striking peculiarities. There are numerous technical terms. Many of them are defined in the text itself, but there are others that are not defined. . . . It is true that we cannot be quite sure of the meaning of many words and expressions in the text. So, too, the interpretation of many passages in the text must remain uncertain. . . .
>
> — Kangle, 1965, Part III, pp. 38-9

'Vocabulary' is the key word; and the characteristic evolvement and entrenchment of this in scientific and technological literatures constitute a special study to comprehend by a modern mind the incomprehensibility of archaic Sanskrit language. Because

> the current content of science can be represented as a system of 'propositions' linked by 'formal inferences' only when we have at our disposal a common, agreed upon vocabulary of terms and concepts to serve as the 'subjects' and 'predicates' of all those propositions; and manifestly, all really profound changes in scientific thought and theory have brought with them correspondingly profound changes in basic vocabulary of scientific terms and concepts.
>
> — Stephen Toulmin, 1977, p. 148

We have to deal with a situation in the history when a language, whether 'symbolic' or 'an ordinary one used in special sense, that is by development of scientific jargon', meant for conveying ideas to be understood by all competent people, was yet to be developed. Without this consciousness, it becomes also difficult to trace and extract the scientific meaning of the works of our predecessors with the mind attuned to the current form of science enumerated above.

Difficulties do not rest solely on the arcane language, but on the manner of treating the blook by scholars. The treatise like *Arthaśāstra* has also not been very critically examined in-depth from the angle of different scientific specialisation, or combined multidisciplinary approach. There had been, of course, a few scattered and discrete attempts, confined to peripheral issues of the dominant theme of the treatise, that is Hindu Polity, leaving gaps in our knowledge in other scientific and technical subjects. This has been felt in many cases. It can be said, and has been said of chemistry, that even a monumental work on this subject by Dr. P.C. Ray did not adequately analyse the basic principles and theories on elements and their compounds, without which their preparation could not perhaps have been accomplished (Das Gupta, 1987, p. 3).

Similar deficiency has also been experienced with regard to Historical Geography. It has been recognised that this subject had also never been joint concern of historians and geographers, and that it was never considered a discipline requiring specialised technique. The result being that study made in isolation from geographical viewpoint only lacked 'historical perspective and inadequate knowledge of source material' (Chattopadhyaya, 1984, p. 2).

So, unless the relevant disciplines are involved, deductions made or results obtained from isolated studies are liable not only to be erroneous, but may diffuse the real worth and rational structure of science in classical texts, as in *Arthaśāstra*.

It has been attempted in the present work to amend these deficiencies by resorting to extensive consultation with experts and referring to writing in relevant subjects in explaining the contents of various *sūtras*, and often for deconstruction of thems (deconstruction meaning 'method of critical analysis applied to literary texts, which, questioning the ability of language to represent reality adequately, asserts that no text can have fixed meaning, and that readers must eradicate all philosophical or other assumptions when approaching a text'). These may be seen demonstrated in various chapters dealing with different subjects of science and technology, and also in the Appendix on this topic. In carrying out deconstruction, enumerations of the devices of reasoning followed by the author of *Arthaśāstra*, as presented in last section XV (*tantrayuktyādhikaraṇa*), have been imbibed as guiding principles. Other substantiating data including those from archaeo-metallurgy, mining, etc., have been incorporated for interpretation of *sūtras* dealing with these subjects.

I was conscious of the snares and pit-falls of more serious nature immanent in such a work, if not carried out objectively, and biases that surface without due attention to appropriate analysis. Two obvious attitudes are observable. One is the Western view of the primitiveness of Indian culture, irrational in nature and worthwhile only when the external influence is apparent in Indian culture, in art, literature, technology and science. Such an identity of India has arisen in West out of diverse perceptions about the country, which assumed a motivated approach during the colonial regime. Overall deduction of these approaches, classified in three predominant categories, viz., exoticist, magisterial and curatorial, 'has been to exaggerate the non-material and arcane aspects of Indian traditions compared to its more rationalistic and analytical elements' (Sen Amartya, 1997). Equal irrationality lies in the perception of many who resort to emotive assessment of worthwhile topics in treatise like *Arthaśāstra*. This attitude mainly dwelling on self-images (called 'internal identities' by Sen) and 'indigenous approach' has generated anti-rationalism in the form of caveat to Western imagination of India and Indian culture. This attitude is also to be derided, as has been done in the manner:

> We look at the sponteneity of the Western world, and seriously ask ourselves, did we even seek truth for its own sake? Had we ever the courage of our own conviction, did we ever follow straight in path dicated by reasons? Had we ever true freedom and sincerity of thoughts and speech or, in brief, had we ever Life? It is indeed difficult to answer these questions, for the popular pseudo-patriotic convictions which exalt at the sight of an aeroplane as being the revival of the *puṣpaka ratha* of the Purāṇas are not only useless but insulting and ludicrous.

As a preventive to unrealistic deductions following any of the type of approaches enumerated, it has been prescribed in very simple terms. That to achieve this one has to 'dive deep into past sanskrit literature with an honest enquiry after truth' (Das Gupta, 1987, p. 51). Like an obedient student, I have tried to follow this instruction, adopting the methods described earlier. It may not be difficult to identify this attempt from the discussions on each of the topics arranged in separate chapters.

A compendium of prevailing ideas about the antiquity and autochthonism of *Arthaśāstra* has been prepared, adding my personal views based on substantiating facts from various disciplines of science and technology. This has been done not for contesting the claim that occidental influence accounts for the brilliance of Indian wisdom in different fields of science and arts. Indigenousness and antiquity of source materials are invaluable and essential for tracing the historical evolution, without which the continuity cannot be tracked. It would leave dark areas in the history, and in turn, breeding ground for speculative deductions.

A number of scholars and researchers considers *Arthaśāstra* as unique in dealing with so many topics displaying keen observation, prudence, ingenuity and originality. Yet they are prepared to attribute such qualities to this book as archetypal only if its antiquity could be viewed as a work of as early as fourth century BC. Winternitz is very candid on this point when he observes that 'It would be the first and the only datable evidence of Indian literature and culture of so earlier an age. And we would have nothing more covetable than to possess such a work' (Winternitz, vol. III, pt. II, pp. 626-7). Attempts have been made to present in the compendium diverse views, facts and findings to avoid any possible anachronism.

I have very consciously tried to follow methods which would obviate the predicaments listed; to make my search for rationality and real meaning of the contents of *Arthaśāstra* in selected branches of science and technology, to conform to 'sure path of science'. But then I am also aware that — *'a paritoṣad vidūṣam na sādhu manye prayogavijānam* (that is, unless the wise and erudites express satisfaction, such methods cannot be taken to have been endorsed as well applied). My efforts would be rewarded only if my work is considered as worth for earning such endorsements.

At the end, a short note is necessary. Throughout the work, every sentence of the original book has been referred as *sūtra;* without going into the structural form of composition-*sūtra* or *śloka*. This constitutes a different subject. But such reference does not interface in any way with the objective or analysis adopted in the present study.

Book, Chapter and *sūtra*, as have been arranged in the translation by Kangle, and which have been used as the source, have been indicated by numerals in consecutive order. For example, reference like 2.8.15 means *sūtra* number 15 contained in Chapter 8 of Book 2.

1

Minerals, Gems, Ores
Identification and Occurrence

Book 2, chapters 11, 12 and 13; Book 4, chapter 1 and Book 7, chapter 12 of *Arthaśāstra* provide vast spectra of information on gems, including pearls; metals and ores, metallurgy and mining. These information are primarily utilitarian, since the produce of the earth, as derived from its surface, have been recognised as of value and contributory to the king's treasury, and hence demanding particular attention. The philosophy in *Arthaśāstra* about treating the natural products with particularity is based on this perception. It is unequivocally stated in *sūtra* 2.6.1 —

> The Administrator should attend to the fort, the country, mines, irrigation works, forests, herds and trade-routes (as the main source of revenue).

The elements which constitute the resources for the purpose of revenue, have been elucidated in the *sūtras* 2.6.2 to 2.6.7, of which *sūtra* 2.6.4 defines the elements under 'mines' as

> Gold, silver, diamonds, gems, pearls, corals, conch-shells, metals, salt and ores derived from the earth, rocks and liquids, — these constitute 'mines'.

All measures are to be taken to ensure a rich treasury,

> For, a king with a small treasury swallows up the citizens and the country people themselves. — 2.1.16

Further

> The treasury has its source in the mines; from the treasury the army comes into being. With the treasury and the army, the earth is obtained with the treasury as its ornament.
>
> — 2.12.37

> Thus from the mines he should collect the price, the share, the surcharge, the monopoly tax, penalty, duty, compensation, fine, inspection fee, and manufacturing charges as well as the twelve kinds of metals and commodities (made from them). In this way he should fix the collection (of income) under various heads in the case of all commodities. — 2.12.35-36

> What is produced from ores, he should put to use in factories for the respective metals. — 2.12.18

> And

> He should establish trade in manufactured goods in a single place, and (lay down) a penalty for those who manufacture, purchase or sell elsewhere. — 2.12.19

The Director of Metals, besides establishing factories for copper, lead tin etc, should also (establish) trade in metal ware.

— 2.12.23

Thus, all forms of mineral-based trade and industries used to be in operation under the governmental agencies, were sources of revenue. Close monitoring of activities from mining to finished products and marketing would ensure earnings and detail knowledge in every sphere of such activities had to be the the stock-in-trade of the king's representatives.

> The knowledge of the mineral world began with the use of stones that were picked up from the ground and directly employed as the most primitive tools of labour and arms. As early as those remote times man was able to differentiate stones and select those that were more appropriate in form,

> mass and strength. Later, with the transition to more improved forms of processing, stones acquired even greater importance since their implementation for practical purposes had undergone substantial changes. Different properties of stones were made better use of, and the knowledge of how to discover and extract them had been gained. — A.V. Milovsky and O.V. Kononov, p. 9

The minerals, metals and ores having been identified as revenue contributing items, it becomes imperative to establish methods and means of ascertaining their intrinsic qualities, knowledge about their origin, source of availability and other related characters contributing to their values — either as 'value in use' or 'value in exchange' as defined by Adam Smith in his *Wealth of Nations*. Those who were vested with the responsibilities to deal with these materials, in mining, in trade, in accepting them in the treasury, must be endowed with these knowledges.

Sūtras quoted in Appendix 2 under the heading of 'Mines and Mineralogy' deal with the concepts and practices about mineral wealth, their identification, classification, methods of concentrating metals from ores, extraction and utilization, etc. These are the knowledge prevalent at the time *Arthaśāstra* was written, in the fourth century BC, which witnessed the maturity in the developmental history of different facets of human civilization, that began in the first millennium BC. It is said of this time that,

> It was then that our intellectual, moral and religious civilization was born and that the foundations were laid on which we continue to build, despite difference in the superstructures we have erected and go on erecting.
>
> — Eric Weil

The *sūtras* are proposed to be scanned in search of the science content latent in them, and the status of this part of the treatise in the context of the modern world of science (Haldar and Sen Sarma). *Sūtras* quoted here may be re-arranged as below:

- Book 2, chapter 11, *sūtras* 2, 3, 4 and 5 — deal with pearls; --

their source-localities, origin and characters indicative of better qualities and defects in them.

- Book 2, chapter 11, *sūtras* 28 to 42 and Book 2, chapter 12, *sūtra* 17 — deal with various precious and semi-precious stones (gems), their source-localities, physical characteristics for identification, like colour, crystallographic shapes, qualities of better varieties of gems, defects.
- Book 2, chapter 12, *sūtras* 2 to 7, 12 to 16; Book 2, chapter 13, *sūtras* 3 to 5 and 10 to 12 — deal with varieties of ores and metals, their identifying characteristics, locations of occurrences, etc.
- Book 2, chapter 13, *sūtras* 17 to 24 — deal particularly with identification and testing of gold.
- Book 2, chapter 12, *sūtras* 8 to 11; Book 2, chapter 13, *sūtras* 6 to 9, 13 and 14 deal with various processes of purification and benefication of metals and ores.
- Book 2, chapter 12, *sūtras* 18, 23, 24; chapter 13, *sūtras* 15, 16, 44 to 56 deal with metallurgy of copper, lead, tin, brass, steel, bronze, iron, bell-metal, etc.
- Book 2, chapter 12, *sūtras* 1, 20 to 22, 35 to 37; Book 4, chapter 1, *sūtras* 51 to 53; Book 7, chapter 12, *sūtras* 13 to 16 deal with mining.
- Book 2, chapter 12, *sūtras* 28 to 34 deal exclusively with salt.

Physical features used for mineral identification

Chapter 11 of Book 2 deals with the 'Examination of the precious articles to be received into the Treasury'. The Superintendent of Treasury is entrusted with the responsibility of receiving jewels, articles of high value, articles of small value, forest produce, etc. The said official is also required to preside over the bureau of experts in various lines, and hence is expected to be well versed in the qualitative characters, trading techniques, assessment, valuation, etc., of the materials to be taken in the Treasury as the wealth of the Ruler. The *sūtras* 2 to 5, 28 to 42 and *sūtra* 17 of chapter 12, Book 2 elucidate the characters, properties, sources, identification, purity, acceptability and values of the gems — mainly pearl, diamond, ruby, beryl, sapphire and coral and semi-precious stones. *Sūtras* 7 to 27 are devoted to the description of different types of strings made up of pearl and often studded with gems.

The above information about various gems has been classified with respect to the following parameters.

1. Identification of minerals/gems based on
   - colour
   - lustre
   - shape and crystal structure
   - specific gravity (light/heavy)
   - hardness.
2. Defects which reduce the value of gems.
3. Source — origin and places of occurrence.

Proper understanding of the modern theories underlying these physical properties recognised and utilised for identification of gems and minerals, as recorded in *Arthaśāstra*, is the primary requirement of acknowledging their scientific basis in terms of modern science.

Gems all over the world are sought mainly for their beauty, either for their colour linked with transparency and brilliance, or for other delightful optical effects. Both these characters in the natural state of gems depend on their crystal structure, which determines transparency, lustre, sheen, display of colours, etc. Artificial cuts often enhance the behaviour of light transmitted through or from the surfaces. These also control the display of colours due to interference, diffraction, refraction and reflection. The properties for which gems are normally adored and sought for have been documented in the *sūtras*, and it is intended here to examine these in the light of the modern knowledge about these properties.

COLOUR

The first obvious and conspicuous property of minerals is the colour.

> Minerals are distinguished by an extraordinary variety of colour and shades, richness and intensity. Some mineral species are characterised by a constant colour, which enables to determine them almost unerroneously. It would suffice to adduce vermilion, golden yellow pyrite, green malachite, blue lazurite. Other minerals, such as tourmaline, beryl, garnet, fluorite, calcite may be of various colours. Thus, for instance, the crystals of tourmaline are found to be black, green, blue, brown, rose or even lacking any colour at all.

In *Arthaśāstra*, such observations are recorded in respect of beryl (2.11.30), sapphire (2.11.31 and 2.11.32), diamond (2.11.39), etc.

> The intensity with which colour manifests itself depends on the aggregate condition and the character of the surface of minerals. In large crystals and coarse-grained aggregates the colour of many minerals is usually darker than in smaller grains, finely granulated or dispersed aggregates. Thus, the outwardly black hematite in earthy aggregates is dark-red . . . golden yellow pyrite in earthy aggregate is black . . . colour of minerals is of various nature . . . the following basic types are singled out (Fersman A. Ye: quoted in Milovsky *op. cit.*, p. 68):
>
> 1. The colour which is wholly conditioned by the characteristic features of the chemical composition, the crystalline structure and the presence of impurity elements — chromophores or the deficiency of minerals (idiochromatism, or colour proper)
>
> 2. The colour associated with the mechanical inclusions of brightly coloured foreign minerals (allochromatism, or the alien colour)
>
> 3. The colour which is the result of such phenomena as the scattering or interference of light (pseudochromatism, or the false colour).

> The colour of minerals has a marked diagnostic importance since it reflects the characteristic features of the structure and composition of minerals. It is employed to determine the content, valency and forms of inclusions of elements — chromophores, as well as in establishing various imperfections in the structure of mineral individuals. Colour also helps to speak of the genetic type of the deposit, temperature or the redox conditions of mineral formation.

Colours of minerals and ores, thus, assume formidable importance in identification, localisation, association, even mode of origin of mineral and ore deposits, that is, as an exploratory feature of various ores. For instance, green and blue selvages of malachite and azurite (ores of copper) help to locate copper ore, brown and rusty colours of hydroxides of iron make it possible to establish the location of oxidation of ore deposits. These points are well documented and highlighted in *sūtras* 2.12.2 to 2.12.7, 2.12.12 to 2.12.17, 2.13.3 to 2.13.5 and 2.13.10 to 2.13.12. Even in modern practice of geology due importance is given to the colour of minerals.

> The colour of minerals should particularly be the concern of the geochemist-prospector and the field-researcher for whom it is the primary feature serving to locate the deposit of mineral resources and what can be of more value than an experienced eye?
>
> — A. Ye. Fersman, quoted in Milovsky pp. 72-3

TRANSPARENCY

Transparency is the degree to which any substance is capable of transmitting light. Thus, a substance is transparent when it is possible to see clearly through the substance. It is translucent when a substance is capable of transmitting light, but no object can be seen through it; an opaque, when no light is transmitted through any substance. Partial character of all these define semi-transparent, semi-translucent and semi-opaque substance. Transparency of a mineral is affected by colour — dark coloured objects are less transparent. Inclusions, that is the presence of foreign material in a crystal, and other internal flaws also affect transparency. Thickness of a substance, say of a mineral sample, also controls the transmission of light through it. In substances having greater thickness, light gets more attenuated and hence the degree of transmission of light through it gets considerably reduced.

The degree of transmission of light through different minerals is different; even in one mineral this varies from one direction to other. This property can be measured by comparing the velocity of light through a mineral v to that in the vacuum c (approximately in the air), and this ratio is called the 'refractive index'. This can be expressed mathematically as

$$n \text{ (refractive index)} = c/v$$

The refractive index varies from mineral to mineral, depending on their crystal structure, and is used often as a distinguishing parameter for identification of mineral species.

LUSTRE

It is the optical effect created by light on getting reflected from the surface of a crystal. Lustre is related to an ability of the surface to reflect an incident ray of light, and is expressed in percentage as the relationship of the intensity of the reflected light I_1 to that of the incident light I_0 or $R=I_1/I_0$ (per cent), and on the refractive index. Since lustre is a function of the intensity of reflected light, this, in other words, depends on the polish of the surface that reflects. In case of minerals, the one which is harder and which has higher refractive index normally attains more polish. Various types of lustre are recognisable by naked eye and each one such results from the interaction of the properties like refractive index, hardness, polish of the surface, intensity of the reflected light, etc. Lustre, thus, indicates also such properties of minerals or ores, and has rightly been identified as their distinctive optical characteristic. Transparent and semi-transparent minerals with a low or moderate refractive index (n = 1.3 to 1.9) and low reflective capability (R = 2-9 per cent) are characterised by glassy lustre. Adamantine lustre is typical of a mineral with higher index of refraction (n = 1.9-3.2) and reflecting ability (R = 9-27 per cent). The refractive index of minerals with metallic lustre is greater than 2.6 and reflecting ability exceeds 20 per cent. On the uneven surface of the crystalline aggregates as well as on the dull surface of the cryptocrystalline and amorphous aggregates the lustre is waxy and dull. Greasy lustre is characteristic of transparent minerals, while resinous is typical of the non-transparent ones. A large number of minerals with fibrous structures have silky lustre. Pearly lustre is typical of scaly minerals, like muscovite (mica). Researches into the nature and variation of lustre in different minerals have indicated that

> lustre depends on the type of chemical bond. Metallic lustre is shared by minerals with metallic and covalent-metallic bond. The diamond lustre is typical of minerals with covalent bond; minerals with ionic type of bond have a glassy lustre. The lustre is the most important diagonstic feature of minerals.

Arthaśāstra has imparted considerable importance to lustre for identification of gems, ores and metals. In *sūtras* 2.11.31, 2.11.33, 2.11.34, 2.12.5, 2.12.17, 2.13.4 and 2.13.14, lustre has been adduced as a character for identification. As in the modern practice, in spite of causality of lustre being correlated to chemical bond or reflection and refraction of minerals, instead of expressing mathematically their natural relationships, the concept of lustre is conveyed through common examples. Thus, the lustre is expressed as 'greasy', 'waxy', 'glassy', 'metallic', etc., and it becomes easy to preceive the nature and character of such lustre. *Sūtras* 2.11.31 can be quoted to demonstrate that such a practice was adopted for conveying the meaning through examples of known and common experience —

> The Sapphire, with blue line, like the *kalaya*-flower, deep blue, having the lustre of the jambu-fruit, having the lustre of the dark cloud. . . .

And such a style of expression was very consciously adopted, for elucidating a scientific subject, as elaborated in the chapter on *tantrayuktiḥ* (Book 15) of *Arthaśāstra* (*sūtra* 15.1.27, 15.1.47).

Aberrations in the physical or physico-chemical properties introduce unfavourable conditions in minerals, particularly in gems. These are defects in gems, the common features being fractures, cleavages, inclusions and intergrowth.

Fractures in minerals may be induced during the process of crystallisation, or may be imposed subsequently due to external stress. Minerals with imperfect cleavages or devoid of them may split into uneven surfaces of the fracture. These fractures may be different from and are characteristic of certain minerals. For example, native sulphur, apatite, cassiterite, etc., have uneven fractures; feldspars have step like fracture, actinolite, tremolite have splintary fracture; quartz, chalcedony, opal have shelly fracture; native elements, like gold, platinum, etc., have haekly fracture. Though fracture in minerals, particularly in precious and semi-precious stones, are responsible for reducing their worth, they are often helpful in identifying mineral species.

Cleavages are planes along which a mineral can be easily split. These planes are normally disposed along preferred crystallographic directions. Cleavages are connected with the crystalline structures of minerals, and hence they function as diagnostic features of minerals.

Inclusions are the presence of extraneous materials other than the chemical ingredients of a mineral in the crystal in visible form. Extraneous materials, gas or water bubbles may be present in a mineral during its crystallisation from the parent melt. Presence of such inclusions often adversely affect the transmitting capability of light through any crystal, in addition to rendering impure appearance.

There are various forms of inclusions of foreign minerals in crystals. Some are trapped during the process of mineral growths forming variegated intergrowths, which usually represent mono crystals with inclusions of mineral grains of enclosing rock.

There are aggregates of minute crystals lining a rock cavity or crystallised gels, disintegrated solid solutions, as well as zoned and sectoral crystals. Heterogeneity is often caused by a process involving a change in the mineral composition, deposition of newly formed minerals in fissures or by replacing them with new minerals.

The structures of minerals produce another effect, called SHEEN, which imparts attractive qualities to them. Optical effects are created by the reflection of light from beneath the surface of any crystal face. The effect created is influenced by the various structural arrangements within the crystal. Thus, 'cat's eye' effect called chatoyance is caused by reflection of light from parallel groups of fibres, crystals or channels; 'asterism' or star effect is caused by a particular arrangement of structures, where three sets of parallel fibres are intersecting at angles of 60° or some other convenient angles. Such phenomenon can be seen in ruby and sapphire. 'Opalescence' reflection pattern gives rise to silvery or pearly appearance; 'adularescence', also known as 'schiller', is a blush sheen oberved in mineral like moonstone, and is caused by thin plates, formed of laminar twinning, within the crystal, interfering with the reflection of light, 'Irridescence' is characterised by the play of rainbow colour, caused by interference of light by a thin surface layer composed of a regular pattern of microscopic spheres.

The optical properties of minerals described above are related to the velocity of light through the minerals. In some minerals light is transmitted with equal velocity in all directions. Such are called isotropic minerals. In anisotropic minerals light is transmitted with unequal velocities in different directions. These include all the crystals except those of isometric system.

> The various wavelengths that make up visible light are reduced unequally in velocity when they pass through a mineral. This gives rise to an unequal refraction of the different wavelengths, and the result is that flashes of vivid colour emerges. This property of many transparent gems is called dispersion or prismatic fire. Diamond, demantoid garnet and synthetic rutile are strongly dispersive. Because of the variable crystal structure — which is the external expression of the internal structure of a mineral, some minerals have the property of double refraction. In some

minerals the two beams are absorbed unequally and emerge as different colours. This effect is known as dichroism. The beauty of colour of gems, such as Ruby and Sapphire is caused in part by, blending of these dichroic colours.

— *Encyclopaedia Americana and Britannica-Micropaedia*

CRYSTAL FORMS

The morphology of minerals, that is their crystal forms, are important in many ways, other than influencing dispersion, absorption, refractive index, reflection, refraction, etc., which contribute to the desirable properties of gems. The crystal forms are useful in distinguishing minerals, different minerals crystallising in different forms. Thus diamond crystallises in the system known as isometric, commonly having octahedral form, that is eight similar faces, each face having the shape of an equilateral triangle. Ruby (*padmarāga* in Sanskrit), sapphire (*indranīla* in Sanskrit) crystallise in hexagonal system, having mainly six faces, each of the shape of a rhomb. Beryl (*vaidūrya* in Sanskrit) has also the hexagonal crystal form, varying from ruby and sapphire, in having six rectangular vertical faces and two-parallel horizontal faces. Emerald (grass green variety), heliodore (honey yellow variety) — all considered as gems, belong to this crystal system.

The atomic structure of minerals, which characterises the crystal forms, also influences the hardness — which may vary more or less in different directions. Hardness, which is also a contributing factor to durability, a desirable charactor of gemstones, is thus determined by crystal system. Crystal structure is also of importance to the lepidiary who wishes to cut a stone to the best advantage. The importance of hardness has been described in *sūtras* 2.11.40 and 2.11.41 as

> Capable of bearing blows, with symmetrical points (capable of) scratching a vessel. . . .

and

> That with points lost, without edges and defective on one side is bad.

It may be noted that expressions like 'symmetrical points' (2.11.40) and without edges (2.11.41) indicate the recognition of crystallinity — symmetrical points expressing the pattern of atomic arrangement in an isometric system (in which diamond crystallises) and 'edges' are defined by the intersection of faces having regular geometric shapes. Non-crystalline minerals normally are not considered as of gem variety (2.11.36).

Though applied only in case of testing purity of gold, *Arthaśāstra* indicates another very important identifying character. This is the concept of 'STREAK' of minerals/ores. The colour of a powdered mineral, obtained by rubbing a mineral on piece of unglazed porcelain (streak plate), varies from mineral to mineral and often varies from the colour of the same mineral. This test in case of gold is done on black flint, or silicified slate or shale called 'touchstone'. Streak of hematite, ore of iron, gives a typical cherry brown colour. The smoothened surface of touchstone is normally used to test the

> purity of fineness of alloys of gold and silver by comparing the streak left on the stone when rubbed by the metal with that made by an alloy of predetermined composition.
>
> — Robert L.Base and Julia Jackson, p. 693

The properties of touchstone in producing characteristic colours of the gold have been discussed in details. Unless the surface of the stone is smooth and even, and pressure applied, when rubbing, is uniform, the streak left on the stone may be misleading.

SIGNIFICANCE OF SPECIFIC GRAVITY

Sūtra 2.12.7 states —

> In case of all ores, when there is increase in heaviness there is increase in metal content.

This apparently inconspicuous statement was ostensibly made out of observations on the comparative and distinguishing characteristics of weights of various ores. Metals occur in nature either in native form, or mostly in combinations of various elements, forming a compound — a mineral — in definite proportions of one with the other. Though an element may be found in various compounds with other elements, their proportion may vary according to their respective affinity and chemical valency. It has been found that the specific gravity of such compounds normally bear direct relationship with the proportion of elements — particularly evidenced in metallic compounds. Increasing specific gravity normally indicates increasing proportionality of metal content in compounds. A few examples would clarify this. This is shown in tabular form (Table 1.1). (It should be remembered that an 'ore' is defined as a solid, naturally occurring metalliferous mineral aggregate, of economic interest, from which one or more valuable metals may be recovered by treatment.)

Table 1.1

Mineral	Compound of	Specific Gravity	Metal Content
Copper			
♦ Chalcopyrite	Copper, Iron, Sulphur	4.1-4.3	34.64% Copper
♦ Bornite	Copper, Iron, Sulphur	5.06-5.08	63.33% Copper
♦ Chalcocite	Copper, Iron, Sulphur	5.5-5.8	79.86% Copper
Iron			
♦ Goethite	Iron, Oxygen, Hydrogen	4.37	62.9% Iron
♦ Hematite	Iron and Oxygen (iron in lesser proportion than in Magnetite)	5.18	69.94% Iron
♦ Magnetite	Iron and Oxygen	5.26	72.36% Iron
Silver			
♦ Native silver		10.5	100% Silver
♦ Agrentite	Silver and Sulphur	7.3	87.06% Silver
Native Gold		19.3	100% Gold
Mercury			
♦ Native mercury		13.6	
♦ Cinnabar	Mercury and Sulphur	8.10	86.2% Mercury
Lead			
♦ Anglesite	Lead, Sulphur, Oxygen	6.2-6.4	68.3% Lead
♦ Cerussite	Lead, Carbon, Oxygen	6.55	77.5% Lead
♦ Galena	Lead and Sulphur	7.4-7.6	86.6% Lead

While the Table 1.1 demonstrates the increasing metal contents in a mineral or compound of elements — associated with the increasing specific gravity — the latter also indicates the higher concentration of such metal-bearing minerals in poly-mineralic aggregates. Metallic ores invariably occur in aggregate with rocks or non-metallic minerals — called 'gangue'. The common gangue minerals and rocks have specific gravity ranging from 1.37 (unconsolidated sand) to 2.98 (some igneous rocks). The specific gravity is much less than those of the metallic minerals, as shown in Table 1.1. Thus, increasing weight of the ore-gangue combination is also an indicator of higher concentration of metallic ores. These are the facts observed, recorded and enunciated almost in the form of a theory, in *sūtra* 2.12.7 and it has been demonstrated in the preceding discourse that this enunciation has the stamp of the advanced knowledge of that time on this aspect.

Because of greater emphasis on pearls and gold in *Arthaśāstra*, detailed discusions on them are preferred.

PEARLS

Though pearl and coral do not satisfy all the norms of a gem, their colour and lustre appeal much to the man's sense of beauty. Pearls and corals, unlike other types of gems, are produced by biological processes, in oysters or shelled molluscs in case of the former. Coral is a hard calcareous substance of various colours growing in the sea. These are mostly calcareous skeletons of Anthozoa and of some Hydrozoa — the marine life that produces it. Historically the name belonged to the Red-Coral, an arborescent species, found in the Red-Sea and Mediterranean. Pink varieties are also available. The shallow waters of many tropical coasts are also characterised by reefs of limestone, comprised of the crumbled skeletal structures of corals.

The highly prized whole pearls are formed by what is known as encystation. These pearls, often called cyst pearls, are formed in perfectly rounded sac within shells, which are believed to be formed when some parasitic worms invade into the shells. The sac, almost perfectly spherical, becomes the site for the formation of pearls. Such are the most adored ones (2.11.5). The formation of pearls in oysters or molluscs, are often not so perfectly shaped, though these have the optical characters similar to those of the cyst pearls, unless formed within a sac, as described. The biological process of secretion of nacre by oysters or molluscs, when some irritants enter the shell, results in the formation of bulges and blisters on the shells. These are removed and used as jewels. And because of their small size, irregular or distorted shape, these pearls are not so valuable as the perfectly round cyst pearls. These irregular shaped pearls are known as baroque (or barroks) (2.11.4).

These are the thumbnail sketch of the modern knowledge about the source, origin and characteristics of pearls and corals. *Arthaśāstra* also mentions the sources of pearls mainly from the lakes and streams, identifies their formation within shells and conch. The shapes rendering them valuable as jewels and inferior quality for the same purpose, have been described with perfect observation, in *sūtra*s 2.11.2 to 2.11.5 and that of coral in *sūtra* 2.11.42.

The physical environment and places of origin of pearls and corals are very significant, not only for their search and supply, but also to understand the processes involved in their formation. Conch and shells abound in water bodies, but not every place is endowed with pearl or coral forming environment. *Arthaśāstra* records names of places from where these are obtained. The geographical locations of these places are necessary to be identified to decide whether the information supplied and locations mentioned, are factual or fabulous, or even traditional.

Sūtra 2.11.2 mentions such place names as Tāmraparṇi, Pāṇḍyaka-Vaṭa, Pāśika, Kula, Cūrṇi, (Mt) Mahendra, Kardama, Srotosi as the localities where from pearls are obtained.

> In both *Rāmāyaṇa* (IV.41.18) and *Arthaśāstra* (II.11) Tamraparni is associated with Pandyakapata (-Kavata) and obviously a south Indian river. — Sethna, p. 373

Tāmraparṇī has been described as a

> river rising in the Malaya mountain. It appears to be the same as Tambaravari of the present day which rises in the

eastern declivity of the Western Ghats, runs through the district of Tinnevelly and falls into Gulf of Manar.

— Apte

The *Rāmāyaṇa* (IV.41.15) also singles out the South-Indian Malaya mountain as the rocky landmark of Tamraparni region. The *Mahabharata* (VI.6.56) does the same and the *Markandeya Purana* too traces the river of this mountain.

— Sethna, p. 373

Sethna further quotes Saletore B.A.:

The Tamraparni of the south was celebrated in Indian history. The Epics and Purāṇas described it. . . . Kalidasa noted it; and it figured in the epigraphs of well-known Karnataka monarchs. . . . The *Brihatsamhita* . . . mentions one distinctive feature of the river, viz., that pearls were obtained in it. — Sethna, p. 374

The *Rāmāyaṇa* (IV.41.17) also associates it with 'the coast adorned with pearls. Pāṇḍya is the name of a country in the extreme south of India, lying to the south-west of Coladeśa. The mountain Malaya and the river Tāmraparṇī fix its position indisputably. It may be identified with modern Tinnevelly. The holy island of Rāmeśvara belonged to this kingdom. Kālidāsa calls the capital of Pāṇḍya-deśa the 'Serpent-town', which is probably the same as Nāgapaṭṭan — 160 miles south of Madras (Apte, p. 1046). The connection of this region in extreme south of India with northern territories is also indicated by associating 'Pāṇḍyakavata' and Tāmraparṇī.

Pandyakạvata indicates the 'door of the Pandyas', the entrance of the Pandya country, implying a close historical connection between this country and the river Tamraparni.

— Sethna, p. 380

About other places Trivedi mentions,

The commentator Bhattasvamin places Pasa near Pataliputra and Kula in Ceylon. The *Periplus* mentions the

> export of pearls in quantity from Muziris, Nelcynda and Bacare, all on the West coast. *Canakya-tika* places Pasika in Pandya country. The rivers except Kula may be on the west side. A variety of pearl, Kardamika, is attributed by the commentator to the river Kardama in Persia, and he also locates two more Srotosiya and Hradiya in the 'barbara country'. But it is more appropriate to consider them as occurring in slime and fresh water and in lake (Hradiya).

The river Kula is near Mayūragrāma in Ceylon; the river Cūrṇī is near the town Muraci in Kerala as interpreted in *Bhāṣāvyākhyāna* in Malayālam and in Sanskrit commentary called *Pratipadapañcikā* by Bhaṭṭasvāmin (quoted in Kangle, part II). Occurrences of gems at Koṭi and Māla (2.11.28), diamond from Sabhārāṣṭra, Tajjamarāṣṭra, Kastīrarāṣṭra, (Mt) Śrīkatanaka, (Mt) Manimanta, Indravana (2.11.37), gold from Jambū river, (Mt) Śatakumbha, Hāṭaka, Venu, Sr̥ṅgasūkti (2.13.3), salt from Indus-land, etc., have been recorded in the *sūtras* indicated within parenthesis. Not all the places have perhaps been correctly identified. Koṭi and Māla, for example, are variously identified. Some commentators place these two in Malaya. They probably present a break-up of Malayakūṭa which Cunningham thought to be a composite name for two contiguous districts of Malaya and Kūṭa. On the other hand, the location of Maḷayakūṭa has differently been explained.

> We know from *Markandeya Purana*, the *Ramayana* and the *Mahabharata* that the Malaya mountain is the source of the Tamraparni river and the main landmark of its region. Hiuen Tsang also refers to the Sandal-producing Malaya mountain as being in Malayakuta. — Sethna, pp. 379-80

In *sūtra* 2.11.49, the occurrence of sandalwood has been associated with Mala. The geographical layout of the river Tāmraparṇī, originating from Malayakūṭa mountain, the confluence of this river with Gulf of Manār, off the Sri-Lanka island, strongly suggests of Malayakūṭa to be in south India — spanning well across present-day Karnataka and Tamil Nadu and Andhra Pradesh. This region is rich in such rocks that contain minerals of gem varieties. Diamond occurrence in Kurnool, Anantpur, Bellāry districts of Andhra Pradesh is well known. Occurrences of pearls and corals in this treatise (2.11.2, 2.11.42) have been indicated as in coastal rivers. That shells producing pearls and corals require certain optimum ambient environment for them to thrive and that this environment is available in certain zones of the oceans and in certain climatic region, are well known today. We find that at the time of Kauṭilya, this was also not unknown.

From the present-day knowledge of the association of gems, minerals and ores with certain specific rock groups and their spatial occurrence, it becomes quite clear that this knowledge was also known in India at the time of Kauṭilya.

As was pointed out earlier, the historical geography can derive much benefit from geology as well while attempting to locate the historical places, in collaboration with history.

Relevant to this subject some other aspects are very significant. These are related to the mode and locations of occurrence of various ores and minerals. It may be noted from the written account of Megasthenes, which many authorities regard as the only authentic written history of the time following the invasion of India by Alexander in 326 BC, that

> The inhabitants . . . having abundant means of subsistence, exceed in consequence the ordinary stature, and are distinguished by their proud bearing. They are also found to be well skilled in the arts, as might be expected of men who inhale a pure air and drink the very finest water. And while the soil bears on its surface all kinds of fruits which are known to cultivation, it has also underground numerous veins of all sorts of metals, for it contains much gold and silver, and copper and iron in no small quantity, and even tin and other metals, which are employed in making articles of use and ornament, as well as implements and accoutrements of war. —Jain, 1972, p. 31

In relation to the occurrence of pearls and corals it has already been shown how their biogenic origin, the locations where they occur and how the perfections and defects in pearls are guided by the mode of their formation, etc., were so keenly observed, known and recorded.

GOLD

In case of the occurrence of gold, it has been indicated that it occurs in three forms, as (i) native gold, that is in pure metal form (*jātarūpa*), (ii) transmuted by means of liquids, and (iii) those produced from mines, meaning, as in *sūtra*s 2.12.1, 2.12.5 and 2.12.12, *prastara-dhātu*, that is occurring within rock bodies. The occurrence of native gold as flakes or nuggets are the known forms. The type of ore described as transmuted by liquids may mean the origin of gold in the hydrothermal process, pertaining to hot water, particularly water of magmatic origin, which is an igneous process, where ore minerals are deposited by precipitation in fractures, faults or open spaces from watery fluids at high temperatures. This mode of deposition associated with hydrothermal activities is interpretable only from associated geological evidences. It is rather inconceivable that *Arthaśāstra* indicated this process as taking place obervably in nature. But in every possibility, the process described in *sūtra*s 2.13.3 and 2.12.2, as of liquid gold flowing through deep crevices and gorge or canyons, describes water containing gold particles flowing through open spaces in rocks and mountains and depositing the particles in those places. This is the process of placer deposit as known in modern geology. The flowing water carries along with it comminuted metal bearing rocks and the metal or rocks having higher specific gravity get settled down where the velocity of flowing water is slackened, as in placid deep pool of water, or in river stretches having comparatively placid flow or low velocity. Such deposits of gold and other heavy minerals are found along the stream channels, in deeper erosional cavities along the streams and similar favourable locations.

Gold ore or gold in native form is mined — when they occur as veins or stringers, obviously filling the open spaces in rocks when the metal was in mobile form, as in the hydrothermal process. Both such deposits are available in India. The placer ore deposits found in streams such as in Subarnarekhā in Bihar, Orissa and Bengal, Subansiri in Assam and in number of other rivers. A section of population, experts in winning gold particles by washing the river sand still earn their livelihood from this practice along these rivers. Occurrence of gold along the foothills of Himālaya from the Sailada river has been described in *Mahābhārata* — this river having been described as flowing between the Meru and Mandāra mountains (Sharma 1983, p. 177). *Arthaśāstra* describes of gold occurring in Jambū river (2.13.3) which may be the river Jamunā — a formidable river flowing past Jambūdvīpa, by which name India used to be referred in ancient literatures, (Roy Choudhury, p. 71).

Gold occurring within rocks, which required mining operation to recover it, were known from many areas in India, nearest to Pāṭaliputra in Singhbhūm, Manbhūm and in Choṭā Nāgpur. In distant south gold occurrences were known from Mysore and adjacent areas. Kolār and Huṭṭi gold fields still have the gold producing mines in India. Gold working in Mysore is of great antiquity. It is quite possible, according to Sir John Marshall, that the gold found in the ancient city of Mohenjo-Dāro might have been obtained from this source; such a supposition would date the industry back to 3000 BC (Pascoe 1965, vol. 1, p 104 quotes Marshal from his *Mohenjo-daro and Indus Civilisation*). Hāṭaka in *sūtra* 2.13.3 is thought to be having the reference of a location. Could this location be connected with Huṭṭi? Regions like Tāmraparṇī, Malayakūṭa, etc., around Huṭṭi and Kolār have prominently figured in *Arthaśāstra* in several occasions. Two significant names are available in *Arthaśāstra* as having such deposits — viz., Śṛṅgisūkti and Hāṭaka. Śṛṅgisūkti can be conceived to be the high peaks of Sūktimat mountain — the mighty range thought to be associated with the people of Vārāṇasī. Strong opinions exist, based on evidences of the origin of some rivers of the region that the Mount Sūktimat lies to the north of Hazārībāgh district of Bihar and extends from Śakti in Rāigarh in Madhya Pradesh to Dalma Hills in Manbhūm district of Bihar. Śṛṅgī or Śṛṅgavat or Triśṛṅga have been located at Uttar Kuru, north of Meru Parvata (mountain) — which is accepted to be located in the Himālaya (Roy Choudhury, pp. 88-120).

Yet it is possible to think of this location within Kharagpur hills to the east of Kiul and north-east of Rājgīr and Gayā in Bihar. Through the central region of this hill mass, the river Mor flows from east to west. It is described in *Rāmāyaṇa* that in the Aṅga kingdom, during the reign of king Lomapāda, who was a close friend of king Daśaratha of Ayodhyā, this region suffered from acute drought for several consecutive years and being advised that the presence of hermit Risyasṛnga would alleviate the situation, the king succeeded in enticing him to come over to this region. When the region received profuse rainfull on his arrival, king Lomapāda married his adopted daughter Śāntā to the hermit. This place subsequently turned to a pilgrimage, though not very widely known. Śṛṅgirikh Temple is still to be found in the Mor river valley.

Further east, adjacent to Bhāgalpur district, the Rājmahal hills hugging Gaṅgā at its northern extremity has the highset peak known as Śṛṅgarṣi. The city of Kajaṅgal, famous at the time of king Harṣavardhana and described by Hiuen Tsang was located near Rājmahal.

This valley and the area around was also famous, being within the kingdom of king Bakul, who it is said, was converted to Buddhism. About three decades back, plenty of idols of Hindu god and goddess were excavated out from this valley during the construction of an irrigation project. It is apparent that the area was under the influence of 'Mahāyāna' branch of Buddhism, which adopted to the worship of idols of lord Buddha and other Hindu gods and goddesses. It is curious to note that Hiuen Tsang described the hill ranges extending westward from Monghyr as 'Hiraṇya Parbat'. Though the reason for attributing the name is not described, the name itself points to some form of activities associated with gold. Gold deposits have been found in Subarnarekhā river valley, which are still being mined. Mount Sūktimat, as per Roy Choudhury, occupies the catchment of Subarnarekhā river, and occurrence of gold, in ore form or with river sand, is confirmed.

This chapter has dealt mainly with the modes of minerals, ores and gems, using several physical characteristics, as coded in the *sūtras*. It has been demonstrated that these criteria and characteristics are remarkably similar to those adopted today, after having inducted them in the science of geology from the advanced branches of physics and chemistry. These are observable in simple statements in the *sūtras*, and they do not require imaginative or ambiguous interpretations to understand them. A table (Table 1.2) is furnished hereafter which will reveal the similarities of the use of criteria in the identification of minerals, ores and gems, as they are practised today and as they used to be practised as far back as the fourth century BC. The properties, used for identification, are undoubtedly identifiable more accurately and quantitatively with modern gadgets, based on the advanced theories and practices in optics and other branches of science. But what makes it remarkable and appreciable about the scientism of *Arthaśāstra* is the identity of parameters described in recognising the quality of precious stones and pearls, and identification of ores, minerals and metals. If not anything else, *Arthaśāstra* testifies the keen observation and modes of classification, which are two most fundamental attributes of natural sciences.

Table 1.2

Sl.	*Properties utilised for identification in modern sciences of geology*	*References of sūtras in Arthaśāstra where these properties are mentioned for similar purpose*
1.	Colour	• Book 2, chapter 11, *sūtras*: 4, 5, 29 to 35, 39, 42 • Book 2, chapter 12, *sūtras*: 2, 5, 6, 12 to 17 • Book 2, chapter 13, *sūtras*: 4, 5, 11
2.	Lustre	• Book 2, chapter 11, *sūtras*: 5, 31, 33, 34 • Book 2, chapter 12, *sūtras*: 5, 17 • Book 2, chapter 13, *sūtras*: 4, 14
3.	Transparency	• Book 2, chapter11, *sūtras*: 30, 31, 33, 40
4.	Shape, including crystal structures	• Book 2, chapter 11, *sūtras*: 4, 5, 32, 33, 36, 39 • Book 2, chapter 12, sūtra: 6
5.	Specific Gravity (Density)	• Book 2, chapter11, *sūtras*: 5, 33 • Book 2, chapter 12, *sūtras*: 7, 12

6.	Hardness (Soft/hard)	• Book 2, chapter 11, *sūtras*: 40, 41
		• Book 2, chapter 12, *sūtras*: 1, 12, 17
		• Book 2, chapter 13, *sūtras*: 4, 11
7.	Streak	• Book 2, chapter 13, *sūtras*: 17, 18, 20

[Please refer Appendix 2 for the full contents of the *sūtras* indicated above.]

2

Prospecting, Mining and Metallurgy

Prospecting (search for ores/minerals)

MINING is resorted to where mineral or ore is found. Ever since the early man recognised these naturally occurring substances in the earth's crust for his use, either for aesthetic purposes, as gems and precious stones, or for purposes of making of ornaments, weapons, etc., that the search for their occurrences assumed an important human activity. In the earliest phase, these were mostly of chance findings. The obvious queries about the ore-deposits, as to how they are formed, where they are formed and when and why they are formed, must have haunted the human mind. If not merely as knowledge-acquiring enquiries, but mainly the motive to possess the useful metals and minerals, must have been the propelling inspiration to understand properly these questions. The primary requirements are the identification of the materials — mineral and ores — and establishment of the factors which would be utilised as guides for locating them. The former has extensively been recorded in the *sūtras*, as described in the previous chapter. As regards the latter, a good amount of information has been provided by indicating the geographic locations where various minerals and ores were found. In addition, the characteristic geomorphic locales which may be examined to look for different minerals/ores have also been indicated. Instance may be cited of *sūtra* 2.12.2 where it is stated that liquid flowing 'inside a hole, a cave, a table land at the foot, a rock-cut cave or a secret dug-out in mountains' — contains gold. *gūḍakhaṭesu antaḥprasyandinaḥ* can actually be translated as flowing in through the depth of crevices, gorge or canyons. The *sūtra* conveys the idea of an aqueous medium flowing through deep open space as containing gold. When viewed in the light of the present-day knowledge of the ore deposits, this resembles two processes involving aqueous medium as discussed earlier.

It is but natural that activities directed to locate the occurrence of minerals, gems and ores, which were considered as valuable commodities for the king's treasury, would be given highest importance, and it is natural that the activities leading to discovery would be given royal support and encouragement. Though it has not been directly spelt out, but from the very plan and pattern of instructive compositions of the *sūtras*, appearing as dictums to be followed by practitioners in the field, it may not be wrong to infer that the purpose of such dictums was to disseminate the knowledge to people, make them conscious and educate them in working knowledge, so that they can go about looking for ores, minerals and gems. In *sūtra* 2.12.2 — it is suggested that the Director of Mines should be conversant with knowledge about metals, their mode of occurrence, their metallurgy, etc., and would take asistance of experts. He should look for such information for locating the occurrence of metals in the old mines. And such information, in addition to the earlier descriptions about the properties of gems and minerals, would comprise indication about old mines, like (i) presence of dross, crucibles (obviously used for smelting), (ii) presence of coal and ashes, (iii) existence of ores in earth, (iv) other form of ores (having relations to aqueous medium), with 'excessive colour' and 'heaviness' and with a strong smell and taste. The last two items demand elaboration in respect of the *sūtra* 2.12.1. The contents of the *sūtra* deal with the instructions to the Director of Mines, about the signs and symptoms, which would indicate the prospective locations for undertaking mining. Indications of the presence of ores are what are to be looked into for any mining venture. In addition to the direct evidence of the presence of artefacts which were deployed for operations connected with mining and smelting, it has been said that the ores with 'excessive colour and heaviness and with a strong smell and taste' (*atyarthavarṇagauravam ugragandharasam*) are also sure indications of the favourable locations for the existence of ores.

The ground is splashed with distinct and bright colours, either due to the presence of colourful minerals, incrustations of coloured chemical compounds or painted with colourful solutions from decomposed ores or minerals. These happen during the process of chemical weathering. Chemical alteration of copper sulphide ores in the presence of minerals or rocks which can neutralise sulphuric acid, leaves malachite (copper carbonate) having bright green colour and azurite (another copper carbonate) with bright blue colour in the oxidised and supergene enriched zones. Their presence, easily recognisable from the bright colours, indicates enrichment of ores in depth. This aspect requires elaboration.

After the formation of ores by primary processes, they undergo chemical weathering where some elements are taken out and some are added. Chemical weathering process involves the interaction of atmospheric air, gases, nitrogen, oxygen, carbon dioxide, water vapours, etc. In the process the compound mineral is broken, oxygen is added (oxidation), substracted (reduction), water is added (hydration), water is lost (dehydration) and similar such processes take place with other chemical elements or compounds.

In ore-geology, the process of oxidation in nature assumes a formidable role, particularly with the metals in combination with sulphur (sulphides). In oxidation zone, sulphides are decomposed, sulphur is converted to sulphate (with addition of oxygen) and the metals are carried away either in solution or otherwise fixed as stable compounds.

The ground peculiarities (topography), ambient hydrological conditions and the types of ores undergoing weathering, determine the metal concentration, their transportation and fixing locales, laterally or vertically from the original location and resulting to products which remain on the ground surface. The most common latter product is limonite — a compound of iron, oxygen and water — having brown, orange brown and reddish colour with mobility to influence colouring of surface rocks and soil. They are derived from iron bearing natural metallic compounds. Their presence in the form of characteristic structures and texture, indicates the presence of metal in the vicinity from which they have been derived. They invariably occur as capping known in geology as 'gossan' over concentrated metallic deposits underneath, derived out of chemical weathering and enrichment, as in copper deposits. Weathering processes also give rise to secondary minerals having catchy bright colours. Examples may be cited of bluish green malachite (a copper ore); nickel that occurs as one of the elements in a mineral containing iron and sulphur (pyrrhotite) normally leaches out from the oxidised outcrop, but may leave traces in the form of apple green stains; bornite — a mineral of copper-iron sulphide — after oxidation, gives rise to prominent orange to orange yellow coloured limonite. There are innumerable such cases of oxidised and chemically transferred products having loud and bright colours.

In the weathering process, certain elements derived from ores and minerals get dissolved in water. Depending on the dissolved elements, the water smells differently — often very strong and fowl. Hydrogen sulphide dissolved in water emits very strong smell — as of a rotten egg. Likewise solution of gases, derived out of chemical process involved in weathering of rocks and ores, impart different odour to water.

Thus, oxidised minerals at the surface are often characterised by their colours. Strong smell of water flowing past such ground indicates what lies beneath the surface. Ground water in a mineralised locality, particularly where sulphides are oxidised, contains metals and sulphates in proportions ranging from traces to so much as to make water undrinkable. The water also gives off strong smell. The copper bearing springs were one of the significant signs, which lead to the early discoveries of famous copper deposits of Northern Rhodesia (McKinstry, p. 215).

These facts are well-established in modern sciences. These facts were also recognised then as is evident from the *sūtra*s quoted.

GEOBOTANICAL AND GEOCHEMICAL PROSPECTING

The second part of *sūtra* 2.12.2 points out that a water body having colours similar to gold and presence of certain kinds of plants, are indicators of the occurrence of gold in the substrate or surrounding localities. These observations are explainable by the modern-time practice of ore search based on the principles of geochemistry and geobotany. Geobotany and bio-geochemical methods have developed into a powerful tool of prospecting for minerals comparatively very recently, though in old folk-lore and legends, selective plants were attributed the distinctions as indicators of different metals in the substrate.

> Biogeochemical methods of exploration depend on the chemical analysis of plants or humus to obtain evidence of mineralisation in the substrate, whereas geobotanical methods involve a visual survey of the mineralisation by means of plant distribution, the presence of indicator plants, or mutational or morphological changes induced by excess of certain elements in the substrate.
>
> — Brooks, 1972, pp. 1-2

The reasons for using vegetation as guide to occurrence of ore depends on (i) the suggestion that metals and other elements may influence the appearance of foliage, (ii) the fact that certain elements play a role in determining the growth and survival of certain species of plants in a given locality, and (iii) on the well founded observation that certain plants are capable of taking up and concentrating elements selectively from soil solutions. It is interesting to note the list of plants which are poisoned by certain elements in the soil, while others, if they do not actually thrive on the same substances, are at least able to tolerate them and thus grow more abundantly where competition is lacking. Kalvero Rankama has listed a number of plant species whose growth and presence indicate the occurrence of various metals (McKinstry, pp. 216-17).

Similarly, water chemistry has been adopted in mineral exploration as yet another very useful method. It is based on the fact that

> . . . migration of metals shows itself in the transfer of their compounds from the source by waters in contact with the bed rock; and frequently in the repeated resolution and re-precipitation of deposited sediments. Hydrogenic chemical elements participate in the water balance of the earth's crust

> and yield hydrated minerals on their precipitation from aqueous solution. Some of the hydrogenic elements are copper, molybdenum, nickel, silver, zinc, cobalt, mercury, chromium, arsenic etc.
>
> — Ginsburg, 1960, pp. 186-215

From what have been conveyed in different *sūtra*s in Book 2, chapters 11 and 12, there are strong suggestions that various directives meant for the Director of Mines, are actually guides to be followed, may be even by common people, in search of ores. As a matter of fact, there is indication in *Arthaśāstra* that private prospectors were encouraged for necessary information on the occurrence of minerals, ores, mines or buried treasures. Such informers were to be rewarded variously depending on whether the informer was a state servant or not. *Sūtra*s 51 to 55 of Book 4, chapter 1 specifies such prizes. The private finder would get to the tune of one-sixth of the find; in case of a state-servant or those hired for the specific purpose of searching for ore-locations, this would be one-twelfth part. That practice of hiring personnel with promise of incentives was in vogue is brought out by the word *bhṛtakaḥ*, which originates from the root *bhṛ*, meaning hired, brought or fetched, as in *sūtra*s 4.1.51 and 9.1.7. These hired personnel must have been those with faculties for identifying and locating occurrences of gems, metals and ores. This would ultimately assist the state authorities to undertake mining operations. The role of private prospectors in this business in modern practice of search for ores is well recognised and acknowledged. In India, an extensive plan for making laymen conscious about minerals was launched by the premier organisation of Geological Survey of India as late as 1959. The objective of the official programme was announced as:

> with a view to supplementing the expanded programme of mineral development the Geological Survey of India has embarked upon an undertaking, the main objective of which is to publicise widely the possible occurrences of minerals in rural areas and to emphasise their national and industrial importance, so that interested people could assist in promoting the search for minerals.
>
> — GSI, 1959

The programme was widely acclaimed as is evident from the Report of the Working Party of Geologists of Economic Commission for Asia and the Far East (ECAFE) in Tokyo in 1960, which observed,

> The experiment undertaken in India of sending officers of the Geological Survey of India to various districts with mineral and ore specimens in order to make the laymen 'mineral conscious' was also noted with interest. Such experiment would not only make the people aware of the usefulness and importance of geological work but might also help in the search for the findings of mineral deposits.
>
> — *GSI*, 1962

In this programme, boxes containing mineral and ore samples along with booklets giving descriptions of minerals, their distinctive features, etc., were distributed, as guides for the purpose.

In the medieval period Severinus preached this sermon —

> Go my sons, buy stout shoes, climb the mountains, search the valleys, the deserts, the sea shores and the deep recesses of the earth. Look for the various kinds of minerals, note their characters and mark their origin. Lastly, buy coal, build furnaces, observe and experiment without ceasing . . . for in this way and in no other will you arrive at a knowledge of the nature and properties of things.
>
> — Adam Frank Dawson, *op.cit.*, p. 210

The guides that are contained in *Arthaśāstra* can be classified as in the Table 2.1:

Table 2.1: Guides for locating ores*

Sl.	*Guiding Features [Modern Concept]*	*Sūtras in the Arthaśāstra where the modern concepts are revealed*
1.	Physiographic guides	• Book 2, chapter 11, *sūtras* 2, 37 • Book 2, chapter 12, *sūtra* 2 • Book 2, chapter 13, *sūtra* 3
2.	Mineralogical guides	
	I. Physical nature of ores, minerals	• Book 2, chapter 11, *sūtras* 2 to 5, 28 to 42
	II. Alteration, mineralisation, oxidation	• Book 2, chapter 12, *sūtras* 1 to 7, 12 to 17 • Book 2, chapter 13, *sūtras* 2 to 4, 10 to 12, 17 to 25, 27
3.	Geochemical, geobotanical	• Book 2, chapter 12, *sūtras* 1, 2
4.	Structural guides (fractures, veins)	• Book 2, chapter 12, *sūtras* 1, 2
5.	Old workings	• Book 2, chapter 12, *sūtras* 1, 2

* McKinstry, *op. cit.*, chapters 8 to 13. *Note*: For the details contained in the *sūtras*, please consult Appendix 2.

Mining

Arthaśāstra does not speak at all about the techniques followed in mining. It is curious in the face of fact that though other features connected with the activities prior to mining or following mining operations, that is, of prospecting and metallurgy, have so elaborately been developed, this aspect remains completely untreated. That mining operations were in existence as early as the chalcolithic period have been definitely established in Rajasthan — at various localities within the Arāvallī hills, Haryana, Bihar and in extensive areas in south India, specially the gold fields near Kolār and Huṭṭi. Evidences of superficial gouging of oxide rich ores have been found in pits often measuring 7 to 8 m in diameter and 3 to 4 m in depth. Heating and quenching of rock mass must have been resorted to for shattering and disintegration of rock mass containing ores. Besides, in six such sites deep mines were noticed.

> All deep mines opened into galleries and narrow tunnels following natural configuration of chalcopyrite (copper ore)

> ore veins at different levels. All these deep galleries were provided with one to one and a half meter diameter ventilation holes at regular intervals.
>
> Samples of timber supports recovered from a gallery at a depth of 120m at Rajpura-Dariba mine in Udaipur district was dated by radio-carbon method at 3120±160 years before present. — Hegde, 1991, p. 15

Timbers found at a depth of 250 feet at Oakleys shaft of Huṭṭi mine indicates C_{14} dates of 1890±70 and 1810±70 years BP. Though this indicates later date than that at in Rajasthan, there are historical arguments to assume that the mining operation must have been going on from much earlier time. Two such indications are relevant, viz., (i) clustering of Neolithic settlements around Maski fields: Maski, Mudgal, Billamrayan Godda, Anandgal, etc., in south India, all within a radius of 5 miles of auriferous zones, (ii) presence of large broken crushing and rubbing stones among the debris of the Wandalli and Gaudur ash mounds, again in the south (Allchin, 1962, p. 209). In *sūtra* 2.12.2, underground mining process seems to have been hinted. In the south-western part of Haryana evidences of ancient mining and smelting activities have been discovered, in the form of old workings for iron and copper, mine wastes, slag dumps, smelting furnaces and cylindrical crucibles (cupolas). C_{14} dating of charcoal pieces found in the cupola dumps assign an age of 2130±130 years before present (years of processing in 1983) (Dey, 1983, pp. 42-50).

It is interesting to note in connection with the ancient gold mining in south India the contents of the *sūtras* 7.12.24 and 7.12.25. The *sūtras* indicate the preference of southern route, running southwards from Gaṅgā valley, to the northern route towards Himavat, because the 'commodities conch-shells, diamonds, rubies, pearls and gold are more plentiful on the southern route' (7.12.24) and because the route to the south has many mines (7.12.25).

Subsequent indications are also there to prove mining activities in these areas. Aśokan inscriptions have been found in five such places within the Maski area and around. The historical evidences, thus, prove large-scale mining activities during Mauryan period, and its beginning or continuity from a time earlier than Aśoka can be inferred without gross violation of historical and other indications.

It is also interesting to look afresh into the descriptions given by Megasthenes about the Gold Digging Ants. The descriptions about the diggers are not only curious, but appears to be hallucinative, though other associated facts as described in his account are so realistic. It is said that the diggers throw up heaps of earth, at the mouth of the mines; the heap they throw up consists of gold the purest and brightest in the world. The mounds are piled up close to each other in regular order like hillocks of gold dust. There used to be pilferage by the neighbouring people and whenever such attempts were made to plunder the mine-products, the diggers would chase after them. Megasthenes goes on to describe about the diggers that 'it hence appears that they understand the worth of gold, and that they will sacrifice their lives rather than part with it'. It is rather difficult to attribute all these, which are so typical human activities, to some creatures like ants. About this, cogent arguments are adduced

> to prove that the 'gold digging ants' were originally neither, as the ancients supposed, real ants, nor, as so many eminent men of learning have supposed, larger animals mistaken for ants on account of their appearance and subterranean habits, but Tibetan miners, whose mode of life and dress was in the remotest antiquity exactly what they are at the present day.
>
> — Ramachandra Jain, pp. 94-7

Description of the dumping of earth at the mine-mouth conforms to the mucking process in underground mining. Miners emerging out of underground working, completely covered with rock dust and slush always display such grotesque appearance that they can evoke any kind of queer description. This might have been the case with Magasthenes, who, in all probability, did not have any experience of mine workers. His account of India — in different domains of politics, social structure, wealth, agriculture, etc., so vividly and realistically recorded, does not really compare with this part — unless analysed in this form. The description reinforces the inference that one form of procuring gold from rocks used to be by underground mining.

Smelting of ores used to be carried out at mine head or near about, and the nature of smelting process (described later) were seen to be practised in India, in Rajasthan, Bihar and many other localities as late as last century. The culture in the process of mining, ore dressing and metallurgical practices seem to have had continuity through the proto-historic period to the modern times. Chemical analysis of the metal artefacts and slag indicate their source to be from known Indian natural deposits, from Rajasthan, from Bihar, etc. It is logical to infer that the progress of metallurgical techniques as associated with recovery of metal-bearing ores from rock mass which contain them, could not have been divorced from the mining operation.

> Craft traditions and technological processes may survive many major cultural, economic or political upheavals. . . . Those crafts and techniques which relate to the peculiar conditions of a region, by utilising its natural products in the form of stone, metal pottery, textiles, etc., always tend to survive.
>
> — Allchin, 1968, p. 274

The reasons for the absence of descriptions on mining technology in *Arthaśāstra* may be due to the fact that more emphasis was given on the economic aspects of the products and their various uses, once these were mined out. This is obvious from the following references, which are found related to mining operations.

1. Inspection of old or new mines, assisted by experts and implements. — 2.12.1
2. Decision to be taken about undertaking mining venture by the State or private agencies. — 2.12.22
3. Penalities to be imposed on persons breaking mining laws. — 2.12.20 and 2.12.21
4. Collection of revenue, tax and penalities. — 2.12.35-6

Sūtra 2.12.1 actually speaks of the process of examining the prospects, before developing it, as in the two step-wise modern process of developing a mine

> A prospect is a potential ore producer which is still in its early stages of development; not until there is enough ore to support a substantial output does the property achieve the dignified status of 'mine'
>
> —McKinstry, p. 428

The ore, in modern technical parlance, means a body of metal-bearing material, which can be treated at a profit. And, hence, the necessity of detailed examination, and the Director of Mines in *Arthaśāstra* was assigned this responsibility to examine the old mines, after locating one such with guiding identification signs. The assistance to be had from the 'workmen skilled in the work' was meant, in all probability, not only to assess the technicalities, but also to act as a valuer, an assessor, to judge the viability of starting the mining operation. That this aspect of assessment, involving economic considerations, besides perhaps the technicalities of mining, was very much an essential part of the duties of the Director of Mines, is revealed in the *sūtra* 2.12.22. As a matter of fact, the absence of such matters as the techniques of mining anywhere in *Arthaśāstra*, but the most prevalent and emphatic issues like tax, revenue, etc., make it more convincing that detailed pre-examination of a mineral/ore prospect was with the purpose of assessing the economic viability of opening a mine.

Sūtra 2.12.22 elaborates this. It has been indicated that the Director of Mines should work on prospects which is less 'burdensome in point of expenses or working' and the problematic ones should either be mined on sharing basis or should be leased out on royalty basis. The move is not to be seen as any attempt to trim the responsibilities of state enterprise, but may indicate the perception about the economic holding of a prospect in terms of the scale and dimension of state enterprise. The private or individual undertaking with comparatively small overhead could cater economically to the smaller requirements and demands. This might have been weighed against bigger enterprise by the state, in deciding as a policy to allow the prospect on lease or working in partnership. Yet, it has also been clearly indicated that mineral rights are vested with the state. Working on such properties without permission, and sale or purchase of mine-products and finished products elsewhere other than the state channels were punishable acts (*sūtras* 2.12.9 and 2.12.21). It is only with the permission of the state that mining could be undertaken by agencies other than the royal establishment. The absolute royal right over the minerals, mines and mine-products is also clearly spelt out in *sūtras* 2.12.35-6. Thus the Director of Mines is to ensure the collection of price, share, surcharge, monopoly tax, penalty, duty, compensation, fine, inspection fee, manufacturing charges as well as metals and commodities made from them. Obviously, these are related to the mining and manufacturing operations by non-royal agencies. The duties of the state employees— Director of Mines and Metals, are fixed otherwise in case of state undertakings as in *sūtras* 2.12.18, 2.12.19 and 2.12.23.

The concept of mineral resources as the monopoly right of the crown is expressed in emphatic terms in *sūtra* 2.12.37 — that

> The treasury has its source in the mines; from the treasury the army comes into being. With the treasury and the army, the earth is obtained with the treasury as its ornament.
>
> — Kangle, pt. II, p. 110

It is seen that while very high degree of acumen is displayed in *Arthaśāstra* in dealing matters with geology as applied to mineralogy (specially their physical properties, identification, distinctiveness, quality, etc.), mineral economics as applied to mining ventures, mining administration and metallurgy (as will be discussed in the remaining part of this chapter), there is remarkable absence of any treatment on the techniques of mining. Perhaps this technology was in the most primitive stage of surface digging and limited depth-wise extension along the payable ore-deposits, not comparable to the level of development of the techniques of ore-treatment, metallurgy, etc., of that or could it be that there was deliberate attempt to withhold the secrets of mining techniques from people so that valuable mineral deposits could not be exploited by anyone other than the state agency?

Metallurgy

The preceding discussions about the science of geology in general and gemmology and ore-geology in particular, have brought out the concept and practice in these branches about their modernity which compare so startingly with the present-day theories and practices in geology. Some authorities believe

> . . . these theories were purely speculative in character with little or no experimental verification, being the result of only systematic and logical thoughts; yet they may be said to stand in good comparison with some of the most recent and advanced scientific ideas of our own time, and to bear the stamp of high intellectual perfection and sublime intuition.
>
> — Roy, 1956, p. 40

The theories in ancient Indian science have been relegated as speculative, albeit their being 'systematic', 'logical' and of 'high intellectual perception', on the basis that these offered no 'experimental verification'.

This aspect, that is whether the elements of experimentation was altogether absent and whether the propositions in *Arthaśāstra* were completely devoid of the experiences from it, can be looked into — proceeding with the modern perception and concept of science.

Much have been said and discussed about what is science and the role of observations, deduction, induction, experiment, etc., in it. It is in fact, experience, which is in the root of all knowledge and

> it is usual to say that the two sources of experience are observation and experiment. When we merely note and record the phenomena, which occur around us in the ordinary course of nature we are said to observe. When we change the course of nature by the intervention of our muscle power, and thus produce unusual combinations and conditions of phenomena, we are said to experiment.

> Herschel justly remarked that we might properly call these two modes of experience passive and active observation. In both cases we must certainly employ our senses to observe, and an experiment differs from a mere observation in the fact that we more or less influence the character of events which we observe. Experiment is thus observation plus alteration of condition.
>
> —Stanley Jevon, *op. cit.*, p. 400

Viewed from this idea, it becomes clear that technology applied in any activity dealing with materials provides the ground to observe the results of experimentation.

> It was in the way of extracting and fashioning materials so that they could be used as tools to satisfy the prime needs of man that first techniques and then science arose. A technique is an individually acquired and socially secured way of doing something, a science is a way of understanding how to do it better
>
> —J.D. Bernal, *Science in History*, p. 47, quoted by D.P. Chattophadhaya

The theories of science, thus evolve out of the observation from technology — and they 'grow only when they are in close and living contanct with the mechanism of production' (Bernal, p. 47). And this conforms to the definition of experiment provided by Jevons quoted earlier. Authorities on the history of science, like Joseph Needham, Benjamin Farrington, Gordon Childe, George Thompson and others in whose writings history of science has become truly scientific — endorse that science is implicit in technology and hence can at best be inadequately understood if not positively misunderstood — without the broader perspective of technology (D.P. Chattapadhyaya, pp. 69-70). Subjects dealing with the material science and technology of mineral processing in *Arthaśāstra* provide ample evidence of the experimental process involved. Archaeological evidences endorse the same, and the best evidence is found in the way metallurgy has been dealt with in the treatise.

The available archaeological evidences may first be analysed as an introduction to the evaluation of the modernity of metallurgical technology as contained in the *sūtras*.

ARCHAEO-CHEMICAL AND ARCHAEO-METALLURGICAL EVIDENCES

Chemical theories and experimental procedures with special reference to the analytical chemistry have been employed in analysing archaeological artefacts and slag for exact identification of metal characters, impurities, etc. This helps not only to identify the metals, but also their provenance, metallurgical processes in vogue in ancient times and skill of the ancient people in this technology. Various advanced techniques are involved in archaeological chemistry, such as chemical, spectrographic, microscopic examination, various dating process, etc. Materials obtained from archaeological excavations along widely spaced sites, and ranging in age from 3300 BC to 100 BC (Pre-Harappan to Northern Black Polished Ware Culture), were subjected to such examinations. Application of archaeo-technology has established different periods of cultures with distinct characteristics based upon the unwritten records in the form of artefacts of stone, clay and metals. At the same time such studies offered the opportunities of comparing and establishing the scientific worth or otherwise of the available written records on the technological processes

> . . . the material relics of extinct societies often prove to be of surprising significance for interpreting certain archaic words and passages than the method of pure philological analysis. . . .

and

> the literary sources in their turn may sometimes be of much help to the archaeologists for understanding their data.
>
> — Chattopadhyaya, p. 68

Investigations of copper objects, copper slag and coins obtained from archaeological excavations, belonging to the period 600 BC to

200 BC have thrown light on some aspects of copper metallurgy. The copper-base objects comprise bronze fragments, copper bangles, copper fragment, copper sheet, etc., obtained mainly from Rājghāṭ excavation. [Information on this subject in the succeeding discussions are taken mainly from Bhardwaj, 1979 unless otherwise stated.] (See Tables at the end of this chapter.)

From the analysis quoted it has been possible to identify (i) the nature of ores used, (ii) the roasting and smelting process, (iii) use of flux, that is, use of materials for assisting the fusion and removal of gangue materials.

> Rajghat slag consists of iron silicate in which smaller amount of basic oxides, viz., CaO, Al_2O_3, MgO are dissolved. The presence of 5.39% of calcium oxide in the slag may indicate deliberate use of lime as a flux. The other oxides (Fe_2O_3, Sio_2, Al_2O_3, MgO) are usually present in the melting charge. 1.05% of CuO in the slag may be indicative of the fact that rich ores of copper were used.
>
> The comparative study of the slag from Rajghat and Ahar reveals that Rajghat has lime (5.39%) and Ahar has none. Rajghat specimen has no MnO and has much lesser percentage of MgO. Copper ores from Bihar and Rajasthan have high content of silica (more than 50%), iron oxide (about 20%) and alumina (about 10%). Lime is present in small amounts (less than 1%).

The presence of excess lime in slag is strongly indicative of its deliberate use. Lime as a flux in copper extraction is well establsihed process now. And it establishes that by about 600 BC to 200 BC this technique became known to the people in this business, as compared to the process prevalent and practised much earlier (2000 BC to 1600 BC).

Metallographic examination, revealing macrofractures and characteristic micro-structures indicates the various techniques of fabrication of copper objects as consisting of hot working and annealing; casting in sand and clay moulds; chill cast and warm stamping; hot working and then cold worked and annealed, etc.

Artefacts belonging to ages from 600 BC to 100 BC and slag going as far back as 1600 BC (slag from Ahār) indicate them to be multi-metallic alloys. It establishes that the technology of alloying was quite prevalent in those days under consideration, also the fact that the use of a number of metals was also known.

Some more evidences are available from eastern India. Artefacts obtained at Pāṇḍurājār Dhībī in West Bengal and Tārāḍih in Bihar from strata recording C_{14} date ranging from 1230 BC to 450 BC shows some interesting metal-mix in them, indicating the metallurgical process adopted. A dish-shaped specimen with iron base records presence of copper, iron, manganese, aluminium, chromium, magnesium, silica and calcium, as determined by spectographic analysis. The metal composition of an iron base sickle shows the following metals (in percentage weight):

> Copper — 0.153; Nickel — 0.128; Chromium — 0.008; Cobalt — 0.111; Carbon — 0.22.

Examination of the object under electron and metallographic microscopes depicts microstructures, which indicate that the sickle had been

> thoroughly forged at sufficiently high temperature so that the slag particles were mostly driven off. Carburisation was done during manufacturing of the tool by subsequent heating and forging.
>
> — De and Chattopadhyaya, 1989

The studies revealed that iron objects were introduced in the period from 1250 BC to 950 BC and maturity in the practice reached around 450 BC, that is early Mauryan period.

Tārāḍih, located near Bodh-Gayā in Bihar, yielded several artefacts belonging to Neolithic-Chalcolithic period (1250 BC) which showed iron-tin-zinc alloy used for the manufacture of these articles, like fish-hooks, bangles, needles, nail-parer, beads, etc. Metal contents in patina (in percentage weight) of these objects, carried out by Atomic

Absorption Spectrometric analysis have been shown in Table 2.2 (Prasad and Chattopadhyaya, 1987-8). Similarly, the metal contents of several other artefacts found in soil strata recording C_{14} date from 1600 BC to 800 BC have been shown in Table 2.3. These objects have been found from different localities in Bihar and West Bengal.

Concentration of tin and zinc in some of these articles indicates that the alloy type was bronze. Use of tin as an ingredient of alloy has been indicated in *sūtras* 2.12.23 and 2.12.24. But it is interesting to note the higher percentage of tin in the alloy, when the occurrence of ores of tin is very sparse and does not occur in association with ores of copper or iron in any known deposits in India. Its presence in the finished products could not have resulted from the melting of such ores and concentration during the metallurgical process. It forces us to infer that tin was consciously and deliberately added as an ingredient of the alloy, and its intrinsic utility as an alloy-making metal was known, as is evident from the *sūtras* mentioned. On the other hand, presence in trace of noble metals — gold, silver, nickel, and others like arsenic, bismuth, aluminium, manganese, etc., in artefacts (Tables 2.2, 2.3 and 2.4) might have been original components of ores, and might have been concentrated during metallurgical process. The original composition of copper ores from Rajasthan and Singhbhūm in Bihar, which are considered to be sources utilised by ancient Indians for manufacture of implements, may confirm this contention (Table 2.5).

METALLURGICAL PRACTICE

Metallurgy comprises two main aspects, viz., (i) extraction of metals from ores or other sources and (ii) application of the metals to industrial or other purposes. In *Arthaśāstra*, the use of metals is seen mostly in crafts of ornament making, minting coins, weapons (Book 2, chapter 18), etc.

Majority of metals occur in combination with other elements. The first step comprises the methods and processes for separating the metal bearing minerals from others. The steps involved in separation consist of (i) Separation of metal-bearing minerals from other minerals and waste materials, called gangue, in a succession

Table 2.2: Chemical compositions of copper-base artefacts (in weight %)

Locality	*Age*	*Residue*	*Cu*	*Sn*	*Pb*	*As*	*Sb*	*Fe*	*Ni*	*Co*	*CaO*	*MgO*	*S*	O_2	*Total*
Rajghat	400 to	0.84	79.40	13.99	0.09	Tr	Tr	2.23	-	+	2.46	0.74	Tr	-	99.75
	200 BC	0.32	99.93	1.82	Tr	-	-	1.09	0.32	-	1.86	0.38	0.48	-	100.20
Rajghat	600 to	-	98.05	-	Tr	+	-	0.37	0.08	-	0.85	0.17	Tr	-	99.52
	400 BC	-	96.38	-	Tr	+	-	0.21	0.08	-	2.31	0.84	0.75	-	100.57
		-	95.45	-	-	0.25	Tr	0.96	0.15	+	1.75	-	0.20	1.24	100
		-	86.90	-	-	+	-	1.50	1.07	+	1.10	-	Tr	0.87	100
Taxila	300 BC to	1.24	96.24	-	-	-	-	0.56	-	-	0.56	Tr	-	1.40	100
	AD 200	0.41	98.40	-	Tr	-	-	0.80	-	-	-	-	Tr	0.39	100
Ayodhyā	200 to	-	94.77	-	Tr	-	-	1.20	-	-	-	-	-	3.55	100
	100 BC														
Mathurā	100 BC	0.54	95.27	-	0.25	+	-	0.85	0.98	-	1.02	-	-	0.56	100

Table 2.3: Chemical analysis of slags from Rājghāṭ (400-200 BC) and Ahār (2000-1600 BC) belonging to two widely different cultural times

Locality	*Percentage Composition of Copper Slags*								
	SiO_2	Al_2O_3	Fe_2O_3	CuO	CaO	MgO	SO_3	Ni	MnO
Rājghāṭ	39.17	12.02	38.72	1.05	5.39	0.62	1.20	Trace	—
Ahār	28.16	5.96	45.32	0.91		3.02	4.15		1.89

Table 2.4: Chemical composition of patina of artefacts found near Bodh-Gayā (1250 BC)

Objects	*Cu*	*Sn*	*Fe*	*Zn*	$\frac{Sn \times 100}{(Cu+Sn+Fe+Zn)}$
Bangle	6.7	0.4	1.8	0.2	4.39
Fish-Hook	12.1	22.2	13.0	0.3	46.63
Needle	6.4	0.4	1.7	0.02	4.69

Table 2.5: Chemical compositions of artefacts (in weight %)

Sl.Locality	*Antiquity*	*Object*	*Cu*	*Ag*	*Au*	*Bi*	*Fe*	*Pb*	*Sb*	*Sn*	*As*	*Zn*	*Ni*	*Mn*	*Al*
1. Maṅgal Koṭ, West Bengal	Chalcolithic (1600-800 BC)	Bangle	Base	0.10	-	-	0.01	0.05	0.10	10.05	0.10	0.5	0.10	0.01	-
2. Sonpur Bihar	Chalcolithic (1600-800 BC)	Spoon	64.89	-	-	-	-	-	-	32.42	-	-	-	-	-
		Bangle	87.98	-	-	-	-	6.985	-	18.87	-	-	-	-	
		Rod	98.43	-	-	-	-	-	-	0.173	-	-	-	-	-
3. Pāṇḍurājār Dhībī West Bengal	Chalcolithic (1600-800 BC)	Earring	Base	0.10	0.005	0.1	0.005	0.5	-	10	0.005	-	< 0.01	-	0.005
4. Bahiri, West Bengal	Chalcolithic (1600-800 BC)	Wire	Base	0.01	-	0.1	-	0.1	0.005	10	0.10	-	-	-	-

Table 2.6: Chemical compositions of Copper Ore from Singhbhum (Bihar) & Khetri (Rajasthan) [Ore constituents in weight %]

	LOCALITIES							
	SINGHBHUM						KHETRI	
CHEMICAL INGREDIENTS	Mosabani Mines	Mosabani Mines	Rakha Mines	Rakha Mines	Rakha Mines	Roam-Sidheswari Mines	Sample I	Sample II
Cu	3.22	2.039	3.04	2.69	1.51	1.32	1.27	1.57
Ni	0.08	0.108	0.092	0.098	0.07	0.072	0.005	0.005
S	4.37	3.30	4.16	3.09	2.24	2.00	3.39	13.22
Co	-	-	0.012	0.014	0.011	0.012	0.01	0.06
Mo	-	-	0.025	0.033	0.005	0.002	0.001	0.0025
Au	-	-	-	-	0.000015	-	Trace	Trace
Ag	-	-	-	-	0.00023	-	0.00006	0.00008
As	-	-	-	0.0003	-	-	-	-

S_n in the ores of Rajasthan has been found to vary between 0.00015 and < 0.001%

Sources:

1. S.C. Sarkar — Geology & Ore Mineralisation of the Singhbhum Copper-Uranium Belt, Eastern India, Jadavpur University, 1984.
2. M.K. Hore *et al* — Exploration of Copper at Satqui, Khetri Copper Belt: Indian Minerals, Volume 30, Number 2, 1976.

of what are called 'dressing operations', or beneficiation; (ii) Metal-bearing minerals are then treated to separate the desired metal from sulphur, oxygen or other elements with which they are combined; (iii) Separation processes fall into three main categories, viz., (a) smelting, whereby the concentrated ore is decomposed and metal released by heat (pyrometallurgy), (b) the desired metal is selected out by treating the ore with solution of salts in water (hydro-metallurgy), for example cyanide treatment for separation of gold, and finally (c) a more up-to-date method of electrolytic technique — in which metals are separated by electrolysing a fused mixture of salts, the best example being the extraction of aluminium and magnesium. Whatever may be the separation process, which vary according to the type of metal and ores, the crude metal obtained by the process requires purification and refining before it can be put to any practical use.

The standard metallurgical processes may be grouped as follows (Mckinstry, pp. 567-79):

Concentration (Physical beneficiation)	• Hand picking • Gravity method • Floatation • Amalgamation
Reduction	• Pyrometallurgical — consisting of smelting and volatilisation • Hydrometallurgical — consisting of cyaniding and other leaching methods
Refining	• Fire refining, chemical refining, electrolytic refining

Of these, the application of electrolytic method is the most modern one, and obviously should not be searched in the old technology.

Barring this, all the other methods listed earlier are found in use in more systematic and refined form even during the time of Kauṭilya, when compared with the earlier proceses from pre-historic period, as found from archaeological evidences and also from written accounts in pre-Kauṭilya literatures.

Accounts related to different components of metallurgy as classified earlier, indicate that the basic principles of utilising the physical properties of ores were well known. Densities, fusibility and the like, described in connection with characteristics of ores are just not only for their identification, but are meant also to discriminate them from other undesirable associated minerals. Archaeological evidences of separation of ores by gravitation method have been provided by Hegde (*op. cit.*, p. 17), even from the chalcolithic period. He provides evidences of crushing of ores and gravity separation on the occurrences of innumerable crushing pits — 30 to 60 cm diameter and 40 to 70 cm deep adjacent to

> smooth, gently inclined rock surfaces, mostly marked with rows of shallow pits 3 to 4 cm in diameter and depth. The finely crushed ore was allowed to slowly flow down the inclined plane and thus, by repeating this process, much of the gangue was effectively separated from copper and iron oxide.

Though such processes were not clearly described in *Arthaśāstra*, the covert indications in several *sūtra*s point out to the continuity of such practices from ancient times. The very fact that emphasis was given on the metallurgical technology of refining, alloying and multiple uses of metals, which follows only when concentrated metals in ores are obtained, indicates that these could not have been achieved without the knowledge of the first step of the processes.

Yet very interesting and revealing records contained in *sūtra*s 2.12.2. and 2.12.3 attract one to examine them in the light of the modern concepts of colloidal and electro-chemistry. The indications emerge from the descriptions of a gold-bearing liquid, which assumes colours of variable shades, while flowing through and past various topographic ground features. These are deep gorge, (*gūḍakhaṭa* may appropriately be translated as 'deep gorge'), a flat land, etc., in a region known to contain gold ores. The liquid has colours like that of *jambū* (blackberry — of the colour of blackish purple), mango or palm fruit (golden yellow), ripe turmeric (again golden brown), feather of a parrot or peacock (bluish or greenish blue), red (of vermilion), viscous and heavy. Gold, in fine particles, in colloidal suspension in water and mixed with other rock or mineral particles imparts various colours to the liquid. The colour of such colloidal suspension depends on the size, shape and refractive index of the disperse phase and in no way on the bulk colour of the substance. Thus, colloidal suspension of gold is red in colour with very fine particles, and blue with coarse gold particles, as distinct from the typical colour of gold. Many insoluble substances, mostly inorganic ones, can be kept in apparently homogeneous solution. For example, gold, silver, iron, etc., which are not normally soluble in water, can occur in this form. These are not solutions in the true sense, but are heterogeneous, which have been proved by experiments. It has also been proved experimentally that if a small amount of solution, capable of transmitting electricity through them (called electrolyte) is added in such a colloidal suspension, it gradually gets turbid and ultimately the dispersed phase (in this case gold, silver, etc., in suspension) separates out as precipitate.

In *sūtra* 2.12.3, it is said that when this liquid is

> thrown in water, they spread on the surface like oil and absorb mud and dirt, capable of transmuting copper and silver up to one hundred (times their own weight).
>
> — Kangle, pt. II, p. 106

The description of the characteristics of 'liquid ore', lead some to interpret this to be petroleum.

> The description agrees to that of petroleum of Asphalt basis as found in Punjab and Baluchistan. Liquids which ooze out from pits, caves, slopes or deep excavations of well known mountains; which have the colour of Jamun fruit (ripe) . . . or Tal fruit (dull black must be that of ripe ones); and when dropped on water spread like oil to which dirt and filth adhere; of piercing smell and taste.
>
> — Roy, 1946, p. 113

Similar views have been put forth much later by quoting the above passage, commenting further that 'a similar material with stronger smell and taste was known to be bitumen or *śilājatu*' (2.12.4). These materials were erroneously believed to contain gold and mentioned as 'metallic ores' or *dhātu* (Biswas and Biswas, 1996, vol. 2, p. 47).

The latter contention of erroneous identification is unacceptable on several grounds. It would do gross injustice to the author of *Arthaśāstra* to attribute erroneous judgement in recognising the characters of the so-called liquid or/and *śilājatu* as ore of gold, after critically examining the descriptive and genetic descriptions of gems, minerals and ores, as contained in the treatise. It has already been demonstrated how the most modern theories of physics and chemistry could be noticed in the methods prescribed for examining these aspects. The two successive *sūtras*, placed one after the other, must have created this confusion. Instead of considering *tatpratirūpakam* as indicating the likeness of the substance, if this word is taken to indicate the process similar to that in the preceding *sūtra*, that is formation by trickling or oozing out (*prasyandina*) of latent or invisible fractures or joints, in the parlance of geology (*bilaguha*), then *śilājatu* assumes an independent description of an exuded substance having strong smell and taste, without having to consider it yet another form of ore of gold. As a matter of fact such mineral is widely known in Nepal by the same name, and is called by the local people as *pathar kā passeo*, or simply *passeo*, sweat of rock. The mineral is found throughout the lower, central and upper hills of Nepal.

> There is a dark bituminous substance used there and is said to be exuded from rocks. It resembles in external character the bituminous alum or ore (called shale) which is said to be found in Sweden and in many coal mines in England.

Further

> it is said to exude in this state from the surface of soft rocks; and sometimes to be dug out of their substance; and from these sources it is collected in considerable quantities

during the cold and dry seasons. The mineral on chemical analysis showed as composed predominantly of sulphate of aluminium, the taste is 'acidulous saline', and is in great demand in medicine and surgery.

— Campbell, 1833, pp. 482-4

When so much of details are recorded in case of minerals and ores, it is surprising that the inflammable character, the most apparent and formidable character of petroleum could not have been observed and recorded in *Arthaśāstra*. The *sūtras* have other explanations to offer.

The preceding *sūtra* (2.12.2) describes the gold-beraing liquid, which assumes various colours and is viscous in nature. In the same *sūtra, rasaḥ kancañika* is also mentioned. The liquid, viscous in nature, with *paṅka-mala* can be interpreted as a turbid one, assuming viscosity due to suspended particles. And when *kañcanika rasaḥ* is thrown on this liquid, it resembles oil spread over water and takes with it suspended materials (*paṅka-mala*) several times more its own weight. The entire process describes of coagulation by addition of little electrolytic material in a suspension. *Kañcanika* (feminine) is described as a plant related to the genus *Premna*. (Monier Monier-Williams, 1963). About 35 species of this genus are found in India. *Premna corymbosa* (Bengali Synonym — Bhūt Bhairavī, ganiārī, and Hindi synonym — Agethā, Ustabunda) — a small deciduous tree or shrub, is met with near the sea from Bombay to Malacca and also in Sylheṭ (in Bangladesh), Sri Lanka, Andaman and Nicobar. Hence, *kañcanika rasaḥ* is the juice of this plant. Perhaps this juice has the property of an electrolytic solution, causing coagulation and precipitation of the disperse phase in suspension. *Paṅka-mala* can be viewed as describing the turbid condition after coagulation by the addition of electrolyte — which subsequently gets precipitated. This interpretation is supported by the comments given in *Śrīmūl* (T. Ganapati Sastri, vol. 1, pp.199-200):

> *tāmrapalasate rūpyapalasate ba kañcanika rasah bedhśāstrabidhinoikapalaḥ kṣipto tatsarbaṅg svarṇikaroti tuktaṅg bhaboti.*

Except that addition (*kṣipto*) of little amount (*ekapalaḥ*) transmutes it into gold, as described in the commentary (which appears to be a process in alchemy), it stands to reason to consider the action of *kañcanika rasaḥ* as that of an electrolyte causing precipitation of the solid phase comprising gold, copper, silver, etc., in suspension, in water.

Kangle reported the concept of alchemy as emanating out of the meaning of *rasaḥ* interpreted by some as meaning mercury and its effect in transmuting copper and silver into gold (Kangle, pt. III, pp. 71-2). He associated the terms *rasaveddha subarṇa paka* with smelting of gold, and *rasaḥ kañcanika* as 'liquid ore, containing gold'. It has been shown earlier that liquid ore (*rasadhātu* in *sūtra* 2.12.1) actually means the ore carried by flowing water of streams in suspension, resulting in placer deposits, which are mixed with fragments of rocks, pebbles, sands, etc., and which even now are worked by panning river sands, separating particles of gold. In Russian translation also *rasavidhā* is rendered the meaning as 'obtained from streams'. This stands to reason and corresponds with the modern knowledge of occurrence of gold and other heavy metals and ores as placer deposits. At least in two *sūtra*s 2.13.33 and 4.1.49, indications are obtained about the practice of gold panning by washers conversant in the process. *Carakapāṅsum dhavakaḥ* in the *sūtra* mean those engaged in washing the sand particles — *Caraka* originating from the root *cara* meaning 'to be engaged', 'occupied' or 'busy with', *pāṅsu*, meaning dust, sand, crumbling soil and *dhabakaḥ* meaning washing. In chapter 13, Book 2, which deals with the various types of ores, their sources, smelting, purification, etc., the meaning as stated above about washing or panning process of sand (if we pick up the sense of flowing or moving, as meant also by the word *dhavaka*, this explains also the sand particles carried by running water) for recovering gold particles carried along with river sand, does not become out of place or does not imply any import of sense.

The treatment of gold ores by mercury (even if the meaning of *rasaḥ* is taken synonymously with mercury) is a known process of amalgamation. In some places, the same meaning may significantly have described this process of amalgamation. But that question will be dealt with later on. From what have so far been discussed, it transpires that instead of relegating the meanings of *sūtras* 2.12.2 and 2.12.3 as dealing with alchemy, there is strong logical indication that the process describes concentration and separation of heavy materials in suspension in water by the application of organic substance derived from plants or shrubs which act as an electrolyte. The processes of extraction and fermentation of juice from barks, tubers and root-stocks, etc., for preparation of liquors, have been recorded in Book 2, chapter 25 (Kangle, pt. II) — which obviously proves that the properties of the juice were also pretty well known. Kauṭilya gives instances of acids in liquid form, e.g., curd, acids prepared from grains and the like (evidently by fermentation). Extraction of oils from linseed, seeds of nimba, seasum, ingudi, etc., has been mentioned (Roy, 1956, p. 56).

Such advancement in the knowledge of organic chemistry, particularly as related to the process of various products, was seen to have been applied in trying them in metal extraction. Uses of cow's urine, animal bile, cow-dung, etc., in assaying of metallic ores are indicated in a number of cases. In some cases, these may appear to be quite quaint, but all the same it indicates that various substances were tried, may be with success in many cases, in metal processing. In this light, the use of *kañcanika rasaḥ* — a product from shrubs/ plants — as an electrolyte, causing coagulation and precipitation of mineral particles in suspension in water, may justifiably be considered as a tested practice in the metallurgical technology of those days. While trying to refute the idea that *rasaḥ* in *Arthaśāstra* means mercury, Kangle describes the role of *kañcanika rasaḥ* in the process as

> the *rasa kancanikah*, it is stated, can transmute copper or silver, when smelted along with them (*tamrarupyayoh satad upari veddharah*). But the detailed description of *kancanika rasah* given in 2.12.2 can by no stretch of imagination be regarded as a description of mercury. It is this process of smelting liquid gold ore with copper or silver which produces the *rasaveddha* type of gold mentioned in 2.13.3.
>
> — Kangle, pt. III, pp. 71-2

This interpretation of 'liquid gold ore' and smelting of copper or silver to produce gold or transmute them to gold, appears to be equally unimaginative, specially when the concepts, practices and processes recorded in *Arthaśāstra* in different aspects of gem-geology, mineralogy, ore-geology and metallurgy are found so advanced and not demonstrably different from the modern counterparts. The process of causing precipitation after coagulation of solid gold, silver and other particles held in suspension in water (in the process of forming alluvial deposits) by *kañcanika rasaḥ* acting as an electrolyte appears to convey well the meaning of *sūtras* 2.12.2 and 2.12.3, when read along with 2.13.3. It is interesting to compare the contents of *sūtra* 2.12.2 with the idea prevalent in the sixteenth century AD regarding the origin of gold. With reference to gold and antimony, Paracelsus says,

> these are to be found in their 'Ens Primum', or first being, in the form of a red liquor or water which is afterwards coagulated and exalted into gold.
>
> — Adams, *op. cit.*, p. 284

In the *sūtras* contained in Book 2, chapters 12 and 13 (Kangle, pt. II, pp. 105-6), various metallurgical processes have been described. Of these, the evidences of the practice of pyrometallurgy, consisting of smelting and volatilisation, hydrometallurgy — consisting of leaching methods, fire and chemical refining, amalgamation and cupellation are to be found. These are the accepted practices in modern metallurgy. Certain terms are being enumerated below for appreciating the discussion on metallurgy.

CUPELLATION

It is the process of separation of gold or silver from impurities by melting the impure metal in a cupel, which is normally a flat, porous dish made of a refractory, or high temperature resistant material, and then passing a blast of hot air on it in a furnace (in modern technology, this furnace is specially fabricated for effective use of blast air and maximising the process of metal concentration). The impurities, like copper, lead, tin and other unwanted metals are oxidised and partly vapourised in the process and partly absorbed in the pores of the cupel. A detailed description of this process is available in *sūtra* 2.13.49 where the metal is heated in a crucible made of earth mixed with powdered bone, or earth mixed with an equal amount of lead, of earth mixed with sand (*śuṣkatutha*), or earth alone and of earth mixed with cow-dung (Kangle, pt. II, p. 114, footnote). It is interesting to note the descriptions of furnaces and cupolas used in ancient site at south-western Haryana for smelting of iron and copper, about 2130 ± 130 years Before Present (BP), recorded earlier in this section. Description of such device in *sūtra* 2.14.23 is comparable. The furnaces excavated from localities north and west of Mahendragarh are made up of ordinary clay and are somewhat pear-shaped, tapering downwards. These are at least 0.6 m high, external and internal diameters at the top being 0.5 and 0.3 m respectively. The furnaces have an inner baked and hardened lining and an outer soft clayey rim. The inside of the furnace is almost 0.2 m deep, with a flat base of 0.2 m diameter.

The cupolas are cylindrical in shape (maximum 31 cm long), tapering at one end (diameters about 12 cm and 8 cm at two ends) and are hollow. The central opening diameter varies from 2.5 to 3.1 cm, the crucible being open at both the ends. These are made up of two layers — the inner one of reddish ferruginous material and the outer layer of greyish calcareous and siliceous matter with plenty of white quartzite and vein quartz fragments. The broader end is much more baked and glazed, shows snooty projection of opening and the opening diameter at this end being larger. It appears that the cupolas used to be placed on the broader end with the longer axis vertical during smelting the ore, and the cooled solidified metal used to be taken out from the bottom end (Dey, 1983, pp. 42-50).

AMALGAMATION

Certain metals or chemicals have the property of forming amalgam with selective metals, for example mercury forms an amalgam with metallic gold and silver. This principle is utilised in the recovery of precious metals by passing a layer of pulp over a table consisting of a plate of silvered copper, which has been coated with mercury. The mercury holds and partly absorbs the particles of precious metals, while gangue and sulphides pass onward. The mercury and gold are later separated from each other by distillation of the amalgam. It is suited only to ores in which gold or silver occurs in native and fairly coarse particle, as in the case of placers and in many oxidised deposits (*sūtra* 2.12.5 may be seen). Gold that is 'locked up' in grains of pyrite or other minerals escapes recovery. As in the modern mining, amalgamation is used as an auxiliary process to recover coarse gold before applying other processes to recover free gold from gold bearing lodes or concentrates from flotation methods (McKinstry, *op. cit.*, p. 574).

FLUX

In metallurgy, any substance introduced in the smelting of ores to promote fluidity and to remove objectionable impurities in the form of slag is known as flux. Limestone (carbonate of calcium in pure form and carbonates of calcium and magnesium in combination) is used as a flux in iron smelting.

SLAG

It is a byproduct formed in smelting and other metallurgical and combustion processes from impurities in the metals or ores being treated (*kitta* in *sūtra* 2.12.1 and *puṭikittaḥ* in *sūtra* 2.14.23). Slags consist mostly of mixed oxides of elements such as silicon, sulphur, phosphorous and aluminium; ash, and products formed in their reactions with furnace linings and fluxing substance, such as limestone. The slag is the end product of smelting to be left out as waste, and its presence at any spot is indicative of old working. The Director of Mines in *Arthaśāstra* was to have the knowledge of it (*sūtra* 2.12.1). Materials and appliances required in metallurgical activities, particularly in roasting and smelting, such as coal, crucible, etc., have been recorded in *sūtras* 2.12.1, 2.13.49, 2.14.23 and 2.15.60.

Various ingredients have been mentioned in metallurgical methods as effective reagents in bringing desired results of metal concentration. Some of these are difficult to account for. But it is clear that these ingredients do possess the characteristic properties like acidity, alkalinity etc. 'Acids in liquid form like curds, acids prepared from grains and the like (evidently by fermentation)' figure in various *sūtras*. Alkalis derived from the ashes of burnt plants, and in natural form of salts are aslo described. Use of flux, in the form of powdered bones was also known. Apart from the apparently quaint idea of use of certain ingredients like cow's urine, animal droppings, milk, etc., the approach that the metal processing techniques require the use of acids, alkalis or neutral chemicals, corroborates the scientific basis of chemistry. Cow-dung is still used as a bleaching re-agent, apart from fuel. Other processes described earlier, like cupellation, amalgamation, roasting, etc., have been mentioned in connection with purification of gold, silver, copper, etc.

ALLOYING

And finally, the process of alloying by admixture of various metals in different proportions has rather elaborately been described.

This process in metallurgy is perhaps one of the path-breaking techniques which has taken the centre-point in modern industrial world. There is hardly any industry — mechanical, electrical or chemical — where metals with special and desired properties are not used. And imparting such special properties has been possible with the knowledge of mixing of metals. Many metals in their pure state are unsuitable for practical use. By mixing two or more metals, alloys can be produced with a huge range of special properties for particular application. Alloys, as per metallurgical definition, is an amalgam of two or more metals mixed for such purposes as improving strength, hardness, heat resistance, etc. The properties of such alloys are different from those of its individual components. Pure gold is too soft to be used in its natural state and for most purposes is alloyed with some other metal. Yellow gold is gold alloyed with copper and white gold is an alloy of gold and silver (compare *sūtras* 2.13.51 and 2.13.52). An alloy does not necessarily consist of equal parts of the metals forming it. The addition of a very small quantity of one metal can completely change the properties of the metal that is in greatest proportion. In short, with the variation of proportion of constituent metals used for making an alloy, its properties are changed. As of today, some 6000 different alloys are now available, by varying the kinds and amounts of metals incorporated in the mixture (*Mind Alive*, vol. II, pp. 2845-8). Most alloys are made by melting the metals together.

How amazingly these techniques and principles underlying them were revealed in *Arthaśāstra*. It is unfortunate that not much artefacts of the time are available for detailed chemical analysis to verify and corroborate the veracity of codes, on proportion of different metals in producing alloys of desired properties. But the fact remains that the principles of mixing metals were very well known, may be by practice and practical use, as evident from the detailed treatment of the subject of alloying. A few chemical analysis of the objects belonging to the period from 600 to 100 BC (please see the table provided in the earlier section of this chapter) indicates the ingredients present in an alloy. The metals present in larger proportion are perhaps the components used for alloying, and trace elements, in all probability, are the original components in the ore, like, for example, nickel, cobalt, antimony, etc. But consciously added metals, not occurring in original ores, in forming alloys, have also been proved in some artefacts, described earlier.

Modernity of metallurgical techniques in *Arthaśāstra* is found in details about practical arts of metallurgy and metal working. Some of the descriptions, veiled in expressions with dubious meaning of words, appear to give idea about quaint practices — which a number of authorities believe to be evidence of the practice of alchemy in Kautilya's time. The chemical arts related to these fields were considered to be empirical. Regarding the first point, that is about alchemy which is considered as having 'shot through with mysticism and mystery making, lacking clear ideas and denying itself clear descriptions' (Hall, 1964, p. 84), it may be said that this aspect really do not figure anywhere in the writings of *Arthaśāstra*. The descriptions are clear, unambiguous and to the point — perhaps the outcome of the practical experience of the craftsmen. The principles, techniques and processes applied to metallurgy and metal working are methodically recorded. And that these were proved to be just not the result of intellectual exercises, but the hardware knowledge gathered out of actual practice. Perhaps because of these, the methods of metallurgy and metal work remained in current use for next several centuries — even to the modern times. The methods and appliances in use were found exactly in similar form in India towards the later part of the nineteenth century (Ball, 1881). Recent work with ancient artefacts and appliances deployed in metallurgy also confirms this. Broken furnaces and broken tuyere found in ancient work sites reveal:

> The interior of tuyere pipes shows bamboo bark impression indicating that they were made by rolling a layer of 1 cm thick clay over 15 cm long, 3 cm diameter bamboo pieces.

It is not without interest here to note that this was also the method that was used for making tuyeres for iron smelting furnaces until about 60 years ago in our country. Verrier Elwin (1942) who studied the Agaria, an iron working community in central India, in 1930s has given a picture of an Agaria woman making tuyere pipes using bamboo pieces (Hegde, p. 19) (compare *sūtra*s 2.13.49 and 2.14.23). It is interesting to note here the practice of gold refining and purification in vogue in Nepal even as late as the earlier part of nineteenth century. (Campbell, 1834, pp. 622-4). This, it will be seen, compares so vividly with similar processes described in *sūtra*s 2.13.5 to 2.13.9. These *sūtra*s describe the processes of purification and the ingredients to be used for the purppse.The processse described consist of (i) smelting with lead, (ii) impregnating (*niśecayet*) with (ash of) stem of banana tree stem, (this ash contains potassium-sodium carbonate, called *kṣāra* in Assam and used as a delicacy in food), *kanda* (garlic or some form of bulbous roots), *vajra* (which might be *kuśa* grass, blossom of sesamum, or inorganic substance like talc, a hydrated silicate of magnesium, or selenite, a sulphate of lime), (iii) smelting with levigated powder (*kalka*) of these with dried cow-dung, and (iv) pounding of the paste on anvil. The processes are to be repeated till final purification is achieved after removal of impurities. Chemical compositions of the ingredients to be used as described in the *sūtra*s reveal the use of carbonate from the calcined banana tree stem, inorganic compounds of talc or selenite, oil for selective collection of sediments, organic compounds produced from garlic, *kuśa* grass, etc. Processes of heating and pulvarisation, accepted physical processes in pyrometallurgy, are clearly described. The processes described in *Arthaśāstra* and the ones being described hereafter would show the strong similarities of both the ingredients and processes.

Campbell states that gold dust used to be imported in Nepal from Tibet.

> The first step of the process is the melting of the dust, when it is cast into uniform plates about the thickness of writing-paper, a little larger than playing cards; in this state it is as impure as when in the original form, unless as is sometimes the case, insoluble and insulated particles of sand are mixed with it, which of course become separated during the melting. The plates, if cast too thick, are beat out to the requisite degree of thinness, and treated in the following manner: the refiner having selected from some ancient ruined building the oldest bricks he can find, and pounding them into fine dust, proceeds to make up his *masālā* or flux, which is of two kinds: in both the old brick dust predominates, and is considered indispensable to the efficacy of the process. One of the mixtures consists merely of brick dust two parts, of common salt one part, intimately mixed up and pounded together; the other and the best one is composed of brick dust two parts, muriate of soda (salt) one part, borax 1/8th part. The plates of gold being previously smeared over with mustard-seed oil, are piled upon one another to the number of 80 or more, between each pair a thin layer of the above cement being placed. Thus treated, the pile of plates laid on a smooth and earthen floor, and covered with a heap of dry cow-dung (*gobar*), which being set fire to, is allowed to burn slowly to ashes, care being taken that but a limited supply of air has access to the fire; to insure this necessary measure, the process is always carried on a close room — by which means the fire is so slow and weak, so that fusion of the metal cannot possibly take place. The usual time occupied by the burning of the heap of cow-dung is about 20 minutes, after which, when cold, the plates are removed and

examined. The colour, softness, and the shade of yellow left on the touchstone being the guides to their degree of purity.

> The process is generally repeated three or four times only, when the ore operated on is tolerably pure, or where only inferior gold is wanted; but is repeated as often as 20 times, when the dust has been of the white or worst kind, or where gold dust is under operation, and the purest gold called *kundun* is required.

In the foot-note, the whiteness has been explained as 'Most probably having a considerable portion of Tellurium (silver?) in combination with gold.

In support of the above article, James Princep, F.R.S., Editor of the *Journal of Asiatic Society*, vol. III, 1834, writes below the above article, after quoting several experiments by the celebrated French Chemist BOUSSINGAUTT, 'whose experiments, have led to a more accurate knowledge of the subject than was before to be met with even in the best works'. After describing in details the experiments, he writes,

> I have dwelt at some length on the above series of experiments, because they afford a beautiful application of scientific inquiry to a rude and practical process which has been handed down and imitated from generation to generation, without the least knowledge of the real action of the materials upon one another; and so apparently simple, that chemists had hitherto neglected to examine it . . . the brick dust for instance answers a double object, first, to decompose the salt, and thus cause the disengagement of the acid gas; and secondly to absorb with the aid of the excess of salt, remaining undecomposed, the chloride of the gold free from fresh action, and to prevent the loss of the silver by evaporation, for the chloride is of itself very volatile. The porous nature of the pounded brick allows the passage and access of the vapours, and thus gives it a preference over unburnt clay for the object in view. Again the wood fuel, or in this country, the cow-dung cake, giving

> abundance of aquous fumes, is indispensable to the process, while the small intensity of its combustion affords a regulated heat so as not to endanger the melting of the gold, and its open texture promotes the circulation of the moisture, through the pile within. . . .

Regarding the second point, that is empiricism in the field of metallurgy, metal working and other related branches discussed so long, the following excerpt very forcefully answers:

> throughout the Middle Ages when the philosopher, with his strongly logical approach to scientific study, was perhaps more remote from the craftsman than at any other time, the very important technological progress of this period was the product of empirical invention, not science. So much was this the case that, in early modern times, Francis Bacon could contrast the solid inventiveness of the crafts with the sterility of philosophy (as he saw it). The success of empiricism in technology was grounds for believing that empiricism would work well in science. Therefore, Bacon said, science should approximate more closely to craftsmanship, and philosophers should not despise the knowledge that miners, fishermen and artisans possess.
>
> — Hall, *op. cit.,* p. 216

Much before, Bernal (1939, pp. 13-18) described the nature of ancient science. He says

> Science as we know it is a relatively late product. It took definite form only in the sixteenth century, but its roots go back to the very beginning of civilisation and even further, to the origin of human society. Modern science has a twofold origin. It derives both from the ordered speculation of the magician, priest or philosopher, and from the practical operation and traditional lore of craftsman. Until now, far more attention has been given to the first aspect of science than to the second, with the result that the whole progress of science has seemed more miraculous than it was in fact. . . .

Perhaps that is why a refractoriness grew in the minds of the intellectuals to accept primitive science as having the characteristic elements for considering it what is understood as 'Science'.

Relegating the practice of dealing of natural mineral and ores, in the historical past, as mysterious and occult science of alchemy, its worth, as it be, were shelved and avoided. There were reasons behind this, some of which have been discussed in the introductory chapter. Bernal (*op. cit.*, p. 18) analyses alchemy *vis-à-vis* chemistry as:

> Alchemy proved as powerful a stimulus to chemistry as astrology has been to the Babylonian astronomy. Chemistry, unlike astronomy and mathematics, is a subject, which can be grasped only as the result of a slowly accumulated tradition of experiments, requiring but little rationalisation in the way of general theories. In fact, the early theories of chemistry added nothing of significance to the implicit ideas lying behind the operation of the early metal workers. *The practical chemist knew what to do when he wanted a certain result, but he could not have known the effective reason why it worked.* (emphasis added)

Even if it is considered that *Arthaśāstra* contained more of empirical concepts — the present-day metallurgy and metal working based on scientific theories and experimentation also prove their veracity. The continuity of the concepts in *Arthaśāstra* through ages to this day — prove their modernity.

The following table, like before, is intended to provide a classified account of the comparative technologies of metallurgy.

Sl. Process and Method	*Enumerative Sūtras*
1. Concentration	• Book 2, chapter 12, *sūtras*: 2&3 read together
	• Book 2, chapter 12, *sūtras*: 20 to 24
2. Pyrometallurgy	• Book 2, chapter 12, *sūtras*: 5, 6, 11
	• Book 2, chapter 13, *sūtras*: 7, 9, 10, 13, 49
3. Hydrometallurgy	• Book 2, chapter 12, *sūtras*: 8 to 11

4. Refining	• Book 2, chapter 12, *sūtras*: 8, 9, 10, 11
	• Book 2, chapter 13, *sūtras*: 6, 7, 8
5. Alloys and metal working (goldsmiths)	• Book 2, chapter12, *sūtras*: 23, 24 • Book 2, chapter 13, *sūtras*: 4, 5, 25 to 27, 37 to 41, 42, 44 to 48, 51 to 56, 58 to 61
	• Book 2, chapter 14, *sūtras*:17, 20 to 43

ADMINISTRATIVE AND BUREAUCRATIC CONTROL

This chapter would not be complete without comment on the measures of control, monitoring, taxing, pecuniary or other kinds of punishment, etc., which have been elaborately recorded as guidelines to the bureaucrats under the king. It will be seen that all these measures can be propounded only on the basis of wide-scale knowledge on the subject.

In the management of wealth, of which minerals, ores and gems are mentioned to be foremost, the concept of infrastructure with various hierarchical levels of administrators is available in Book 2, comprising 36 chapters, (Kangle, pt. II), of which several chapters deal with the activity of the Heads of Departments. Monopoly right of the monarch over natural resources has been the central theme of *Arthaśāstra*.

The king needs to have monopoly right over the natural wealth, for ensuring his strength, power and authority. Revenue to be extracted requires also to be strictly monitored. Kauṭilya's fiscal acumen, and very sharp one, can be gauged from his identifying salt as a taxable commodity. Salt, the common mineral required by everyone for daily consumption, is a sure and constant source of revenue. Though it is a low-cost mineral, its mass consumption puts this in unique position of revenue earning source. Salt has been given a separate entity — to be dealt by Salt Commissioner. [This compares so well with the happenings in India during the decade of 30s of 20th century, related to the government's control on manufacturing, trade and tax on salt and the extensive agitation against these anti-people measures throughout the country under the leadership of Mahatma Gandhi. This will ever be written boldly in the history of freedom movement in India against British rule.] The full control over them, particularly of the minerals and metals, the manner of trading, imposition of penalty and punishment on violators of codes and stipulations, etc., have category-wise been outlined in *sūtras* 2.12.19 to 2.12.37, in case of mines, metals, and salt. In chapters 13 and 14 of Book 2, these points have been recorded for artisans working with metals — and how meticulously they have been noted about possible pilferage of noble metals from working-spot, possible ways of fraudulent practices by goldsmith, etc., and ways and means to monitor, control and prevent these.

The modern concepts of management, managerial responsibilities, duties, disciplines, systemic approach to the management of mineral wealth can no way be differently conceived than what have been indicated in *Arthaśāstra*.

3

Town Planning, Civil and Architectural Engineering

ARTHAŚĀSTRA, considered as a manual of statecraft and administration contains within it a wide spectrum of technological details, by way of infrastructural adjunct to the various spheres of administration. It may be surmisd that the mind which composed the manual dealing with such a perfect system of administration articulated towards the stability of the kingdom, the prosperity and protection of the king and his subjects,[1] means and measures of abating any external danger, injecting contentment and sense of security in the overall social fabric, would conceive that a strong superstructure can be built only on strong infrastructure. And the latter comprises the suitable dwelling place, buildings, roads, communication, means of production, protective measures, etc. Various branches of Civil Engineering, as we understand today that technology, like town and country planning, layout plans for buildings — architecture, design, construction materials, sanitation, irrigation, etc., are to be found as occupying appropriate places in *Arthaśāstra*. The classification of these and the *sūtra*s which define them are provided in Table 3.1, before discussing in details their worth in the light of the modern knowledge of Civil Engineering.

In *Arthaśāstra*, the *sūtras*, which are related to *vāstu-vidyā*, that is to the more comprehensive technology of Civil Engineering of today, deal with the subjects enumerated briefly in Table 3.1. The

---

1. *prājā-sukhe-sukham rājñaḥ prajānām ca hite hitam . . . nātma-priyam hitam rājñaḥ prajānām tu priyam hitam* (1.19.34).

definition and evolution of *vāstu-vidyā* are relevant in the discussion of the various branches dealt with in *Arthaśāstra*, and as have been clubbed together in Table 3.1.

> *Vastu vidya* was a branch of more comprehensive *śilpaśāstra*; but gradually with time the two became identical. *Vastu-vidya* means the science dealing with *vastu*. The word *vastu* occurs in *Ṛgveda* in the sense of the building site or the building itself. *Vastu vidya* therefore primarily meant the science of Architecture. Later on, however, various other subjects connected to it were incorporated in it. Thus, *vastu*, a place where men dwell includes not only building site and *prasada* or temples and palaces but also conveyances. The *Silpasastra* which included the *vastu vidya* is defined in Sukraniti as a science dealing with public works such as temples, images of Gods, gardens, houses and tanks' (Bhattacharya, 1994 pp. 64).

Public works including sanitation, roads, irrigation, etc., are now considered within the domain of Civil Engineering. And these are the subjects and branches, which find place in extensive discourse in *Arthaśāstra*.

Table 3.1

Sl.	*Elements of Civil Engineering*	*Sūtras Enumerating the Elements*
1.	Town and Country Planning	• Book 2, chapter 1, *sūtras*: 1 to 5 • Book 2, chapter 2, *sūtras*: 3 to 5 • Book 2, chapter 3, *sūtras*: 1 to 5, 15 • Book 2, chapter 4, *sūtras*: 1 to 28
2.	Layout of Buildings	• Book 1, chapter 20, *sūtras*: 1, 2, 10 to 13 • Book 2, chapter 3, *sūtras*: 10, 11, 16 to 35 • Book 2, chapter 4, *sūtra*: 2 • Book 2, chapter 5, *sūtras*: 2, 3, 5, 6 • Book 3, chapter 8, *sūtras*: 13 to 16, 19
3.	Design aspects, including foundation, excavation, stabilityof excavated slopes,	• Book 2, chapter 3 *sūtras*: 4 to 8, 10, 12, 13, 16, 17, 18, 22, 23 • Book 2, chapter 5, *sūtras*: 2, 5

construction materials	• Book 3, chapter 8, *sūtras*: 14, 15, 19
4. Irrigation engineering including hydrology and hydrometeorology, agriculture horticulture (this part has been dealt in a separate chapter)	• Book 2, Chapter 1, *sūtra*: 20 • Book 2, chapter 4, *sūtra*: 5 • Book 2, chapter 24, *sūtras*: 1 to 33 • Book 3, chapter 9, *sūtras*: 27 to 38 • Book 3, chapter 10, *sūtras*: 1 to 3

[Please see Appendix 2 for full version of the *sūtras*.]

Various *sūtras* in Book 2, chapters 1 to 4, discuss the details of town and country planning. The essence of such planning can be classed as:

(a) The settlements might either be renovation of earlier settlements or a new one.

(b) Settlements are to be established in groups of eight hundred, four hundred, two hundred and ten villages. Towns with hierarchical status are to be located within each of such clusters. The villages should comprise of minimum one hundred and maximum five hundre śūdra families of agriculturists.

(c) Such families may eithr be brought from outside the country or from more crowded areas within the country (obviously to avoid a situation where settlements are overpopulated).

(d) The settlements should have well defined boundaries, either in the form of natural physical entities like river, mountain, forests or could be built up, like embankments or planted trees.

(e) The boundaries and frontiers should be well protected by fortresses of various types depending on the physical nature of the countryside in addition to being guarded by identified communities of the society, by ramparts provided with gates at suitable locations, towers for observation and provided with spaces for fighting outside enemies, moats all round.

(f) The settlements to have people of learning and those in different professions to be favoured with land.

Town planning

The manner in which the towns are to be laid out has been outlined in details (Book 2, chapter 4) as under: The site selected on the basis of suitability for accommodating peoples of four castes for laying the town is to be divided into various sectors by 48 ft (8 *daṇḍas*) wide royal highways; three in each direction, should traverse the area from east to west and north to south. The twelve terminal ends of the roads at the boundaries, should have gates at each location, provided with appropriate arrangements for drainage-way and canopied passage. The widths of other types of roads, like paths for irrigation work, paths for elephants, cattle paths, etc., and those running through different types of towns have been specified according to their purpose, importance, and serviceability.

The township is to have the royal palace at the centre, and inhabitants belonging to four castes should have their respective localities. Different sectors are to be alloted for people of various professions — merchants, traders, market-places for different commodities, places for worship, army quarters, physicians, etc. All forms of conceivable infrastructural support systems for civil, military, trade and industries have been assigned suitable locations within the township.

It is amazing to note how meticulously every aspects of social life have been considered for arrangement in a locality! These aspects of the town planning, capsulated in a few *sūtras* form a subject of independent study. How such schemes conform to reality can be visualised from archaeological evidences available so far.

A number of cities located within the Gangetic province and peripheral areas have archaeologically been identified and according to the areas enclosed by the ramparts, these cities can also be graded (Map No. 1). These cities extend to the west into Doab of Ganges and Yamunā rivers, and to the east beyond the Gaṅgā into the complex river system of the Gaṅgā-Brahmaputra deltas. Among the cities of this province are Rājagṛha (Rājgīr), the earliest capital of Magadha; Kauśāmbī, Mathurā, Kapilvastu, Śrāvasti, etc., on the northern edge of the Gangetic plains, Balirājgarh and Bāngarh in

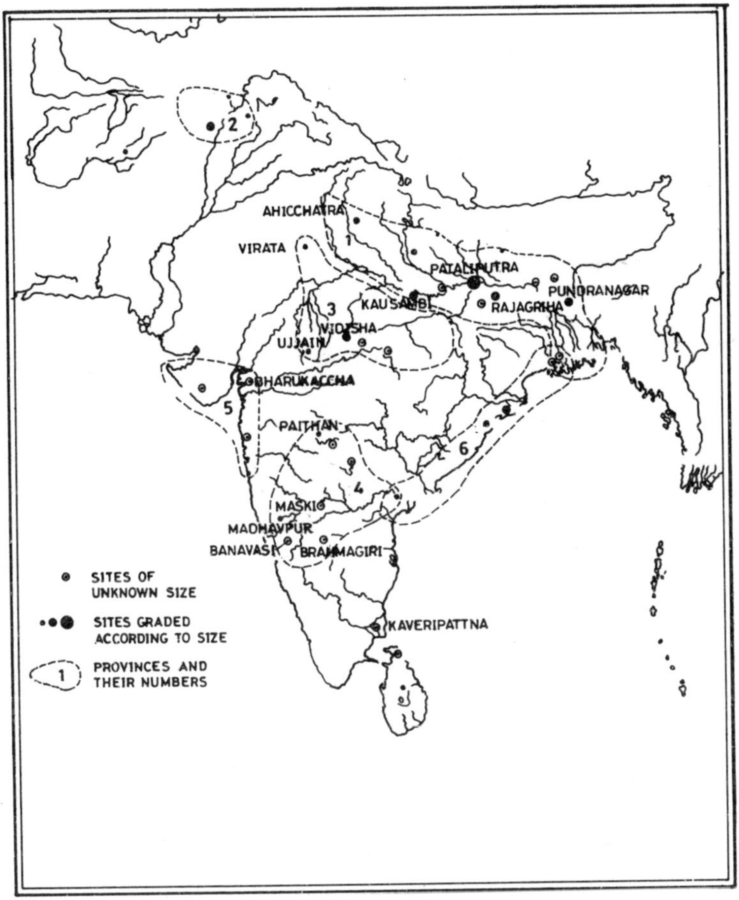

Map 1: Map of Probable Provincial Groupings of the Mauryan Empire, with Cities graded according to their size (courtesy: Allchin, 1994/P-208)

the north-east, and Mahāsthāngarh (now in Bangladesh) in the east. In addition, there are other ancient cities whose size or age have not been definitely ascertained, like Benāras, Campā, Ayodhyā, Vaiśālī, etc. Cities considerably of smaller size situated in Afghanistan and Pakistan, like Taxilā, Kaṇḍāhār, in ancient Gāndhāra, etc., bear the signs of Mauryan characters. Examples of other eminent cities are in Madhya Pradesh, like Ujjainī, western coastal area, like Jūnāgarh, Tāmralipta and Candraketurgarh in West Bengal and Śiśupālgarh in Orissa, and several localities in south India. It is the considered view of the archaeologists that

> there are without doubt many more sites which still await either discovery or proper mapping and identification. . . .

Cities surrounded by defensive walls, moats and ramparts are some of the typological characteristics of the cities, some of which belong to Mauryan period and after. It is nearly true of almost all Gangetic cities. Where a city was constructed on rocky terrain, as was the case with old Rājagṛha, the original capital of Magadha, a solid stone wall with regular square bastions, built on rock, formed a double line of defence around the older city. The neighbouring fortification of the New Rājgīr on the other hand was built on open ground and had a regular moat and rampart.

> Perhaps a better criterion of classifying cities is in terms of their outline plans. We would expect from *Arthaśāstra*, a city to be square with three gates along each side, and roads dividing it into regular square blocks. In the standard *Arthaśāstra* city — there is a central square in the middle of the city, with the royal palace on its north side. The square plan is nearly represented at Śiśupālgarh, where four sides make a near-square, and at Kauśāmbī where there are three sides suggestive of a regular rectangular plan, although it is irregular on the side bounded by the Yamunā river. . . .
>
> — Allchin, 1995, pp. 204-9

The concept and plan of cities, as described in *Arthaśāstra*, have been brilliantly illustrated by Rangarajan (Figs. 14 and 15 in Rangarajan, 1992).

Municipal ordinances to be enforced by the City Superintendent and his staff have been recorded in minute details, with the severity of punishment to be meted out to violators of such ordinances. Sanitation, hygiene, physical environment and all necessary for sustenance of healthy life have been dealt with. And these reveal the mindest of the composer. The dwelling place has not been conceived as a physical entity only — but an expression of a mind — conscious about the complexities and intricacies of interaction between man and nature and social role of a man. Town planning in the context of human civilisation perhaps exhibits the organised mind — in the evolutionary stage of civilisation.

And the organised mind is very well demonstrated in *Arthaśāstra* the way the layout of a town has been planned. In the monarchial system of that time the entire administrative activity was directed to the service of the king whose motto was to be the overall welfare of his subjects, because on that depends the happiness of the king (*sūtra* 1.19.34). The king had to offer adequate protection to his subjects, provide opportunities for trade and business, encourage learning, support productive endeavours, etc. All these are manifest the way space was to be demarcated zone-wise for each type of activity.

Enormous importance is given to the form of a city. It is said that 'cities are product of time', that 'the city records the attitude of a culture and an epoch to the fundamental facts of its existence', that

> it represents the maximum possibility of humanising the natural environment and of naturalising the human heritage; it gives a cultural shape to the first, and it externalises, in permanent collective forms, the second.
>
> — Lewis Mumford, 1944, pp. 4-6

It has further been stated that

> The city is a related collection of primary groups and purposive associations: the first, like family and neighbourhood, are common to all communities, while the second are especially characteristic of the city life. These varied

> groups support themselves through economic organisations that are likewise of a more or less corporate, or at least publicly regulated, character; and they are all housed in permanent structures, within a relatively limited area. The essential physical means of a city's existence are the fixed site, the durable shelter, the permanent facilities for assembly, interchange, and storage; the essential social means are the social division of labour, which serves not merely the economic life but the cultural processes.
>
> — Mumford, 1944, p. 480

Modern analyses of the city structures, as these, are glaringly represented in the town planning contained in *Arthaśāstra*. As a matter of fact the latter seems to have been taken as the model in the characterisation of a town. Interestingly, this medieval period of history marks the advent of town planning in the then civilised centres of the Western world. In Greece fourth century BC was the period which assumed great importance in the history of architecture.

> The development of town planning, and a greater variety of specifically designed civic buildings, are two features of the age. New towns, laid out to the regular plan associated with the name of Hippodamus of Miletus, would be equipped with a regularly planned market-place surrounded by colonnades, and meeting places for the assembly and council. Much greater attention was now paid to the amenities of life, well paved streets with efficient drainage became a normal feature of Greek towns.
>
> — Donald, 1965, pp. 74-5

The Greek world during the Hellenistic period (from Alexander down to 31 BC)

> Saw the foundation of many towns and cities over a wide area, and all the varied aspects of city life were now the concern of the architects and planners. The cities were laid out in regular plans in accordance with rules of town planning developed in the fifth century BC . . . houses were

> laid out in regularly planned areas defined by crossing streets; attention was paid to public hygiene and the comforts of life . . . the varied activities of public life were catered for by well-designed buildings of permanent character . . . the temples of gods were still built on grand scale, and still dominated the layout of the cities. . . .
>
> — Donald, 1965, pp. 101-2

Town planning prevalent in the ancient Western world bears strong similarity with that described in *Arthaśāstra*. The questions of import or mutual intercourse between the two far flung civilisations, through Alexander's invasion have previously been dealt in details. But there is strong logic in the assumption that the internal system of human society and its interaction with nature were influenced by the physical environment, which in turn was modified to the best advantage and use by man. Just like multicentric evolution of organic world, cultural evolution can also be perceived to have originated simultaneously in far and wide centres. The ubiquity of man's approach to nature, the manner of use of raw materials, as analysed from the artefacts lefts behind, might have had its origin from man's perception of nature, propelled by fundamental instincts for survival, protection, shelter and later on, of a complex society. In every stage of evolution, the noticeable factor, which took shape in every sphere of life, was the concept of order. This is manifested in the arrangement of space for living. Though this has been modified to great extent, the 'order' as the basic factor can still be found in the concept of modern towns and cities. In that sense it may be said that in the concept of town planning, the most characteristic feature of the ancient forms has still been preserved even today.

The three models, amongst many others world over, associated with the Chicago school, are worth mentioning, even as ideas pertaining to urban planning and growth. These are in brief:

(a) *Concentric zonal growth model*: In this a city is presented as a set of circles, each one ringing the small one within it. At the centre is the business district, outwardly spreading successively as the centre of wholesale business and light industry, followed by a zone of transition. Next is a middle class residential zone. These in turn are encircled by an outer ring of upper residential zone.

(b) *Sector model*: In this concept some cities grow not in ever expanding circles, but in segments or sectors. Often the growth pattern follows the topography of the land or most favoured transportation routes, whether rail, bus or truck. A river cutting through a city may encourage development of an industrial sector that stretches for miles through the middle of the town.

(c) *Multiple-nuclei model*: Certain kinds of land use are found in separate district areas. Industries are located near the transportation routes needed for shipments of raw materials and finished goods. In a harbour city, for example, industrial areas would be concentrated near the port. Similar uses of land cluster together because proximity helps them — as in financial or shopping districts. Still other land users are motivated by the need of distance. While the wealthy may choose residential areas that command the best views, they also want to get as far away as possible from the pollution and grim of the industrial district and the congestion and bustle of the business district.

— Peter Rose *et al.*, 1982, p. 365

The concepts of city as above brings out a strong similarity with Kauṭilya's concept of physical set-up, specifically of the distribution of city slots according to the classes of occupation in segmented sectors and infrastructural supports for functional integration.

From the dawn of civilisation, when community living was forged into human society, the nature of physical form of dwelling places became a dominant issue of man's intellectual occupation. Several factors influenced this form. Sheltered habitations with closeted economic activities decided this, later on cultural activities became way of town life. Administrative and legal functions became extremely important at one point. Though small by modern standard, the towns, which displayed orderly physical form supporting the different social activities, generated the normative philosophy of town planning. The improvements of transport systems, extreme industrial growth, complexities of trade and social relations, and many other facets of material consumptive society complicated form of city, city life and city growth in modern world. Overcharged with the confronting and coflicting situations making a city far from desirable habitation, the emerging philosophy of city planning is turning to nature and symbiotic relations of man and environment — a concept which is so vividly exhibited in the town planning contained in *Arthaśāstra*. If not anything, the concept of the town according to this treatise certainly supports what Aristotle thought of cities as 'Man come together in cities, in order to live: they remain together in cities, in order to live a good life'. As it is in politics, so it is in city building and arachitecture. The purpose the people have in living together defines the civil form they will find useful and the civil form defines what is required of architectural and urban form. We find the similarity of this concept in the early Romans. They linked a particular form of political life to a particular place called 'urbs'. This produced the concept of city as a legal, civic entity and identification of each particular entity with a particular urban locale.

> The Romans brought the idea of the city down from the imagination of the philosophers and rendered it in brick and marble.
>
> — Van Pelt, 1991, pp. 46-7

Cities and towns are the physical deconstruction of mental concepts and perceptions. In this sense Kauṭilya demonstrates the highest level of concept and perception as revealed in his planning of the city and town.

Architecture and civil engineering

All aspects connected with the practice of civil engineering and architecture of today, especially with the construction of buildings, towers, ground excavation meant either as a protective moat around township or canals, etc., are to be found recorded in code of practice in various *sūtras* of *Arthaśāstra*. These *sūtras* are scattered in chapters 20 of Book 1; chapters 3, 4 and 5 of Book 2; chapter 8 of Book 3.

As for the dwellings, it starts from the selection of site suitable for constructing buildings, location of sites within a town according to the different class of peoples and professionals, etc. It is stipulated in *Arthaśāstra* that the site, on being recommended by experts, should have the boundary fixed by pillars on four corners, connected by iron wires, so as to indicate the extent of the land. The house to be constructed should be in conformity with the extent of the land. If the land is contiguous to another building, the boundary should be fixed at least 3 feet (two *aratni*) away from the neighbour's house. To avoid any nuisance to the neighbours, wells, dung-hills and water course (drains) are to be suitably emplaced between the houses. Similarly, places for carts, pet animals, fire place, grinding mills, water reservoir tanks, etc., should be located well away from the neighbouring house, or at mutually agreed suitable locations in between the houses. All these measures are to be strictly adhered to, for defaulting of which penalties have been stipulated (3.8.3 to 3.8.26).

While so much have been stipulated for the dwelling places, obviously for the common people, the design details and architectural aspects of buildings are described extensively in connection with the royal residence (Book 1, chapter 20) and as related to the construction of forts (Book 2, chapter 3).

SOIL MECHANICS AND FOUNDATION

As are the conventions and practices with building construction, the preparation of foundation has been taken up at the outset (2.5.2). Though this *sūtra* deals with the construction of underground chamber, specification as to how the floor and sides are to be constructed has also been laid down. It states that after a quadrangular excavation is carried out in soil, this has to be dewatered to make it dry. The sides and the bottom are to be paved with big stone slabs, before constructing the chamber with timber, and raising the upper floors. The process of paving the bare excavated walls and floor with stones is obviously meant for insulating the underground soil moisture. But it may be logically inferred that the process is similar to the construction of 'structural foundation' of the modern-day civil engineering. As will be seen later on, such an attribute to the concept is not altogether out of context, when it is seen that *sūtra*s elsewhere deal with minor details about construction within or on natural soils. In other words, soil parameters which come into play in such cases were well known, and in that context it seems quite relevant to interpret that the necessity of providing appropriate foundation to multi-storeyed buildings must have been done with full knowledge of behaviour of natural soil under imparted load. The principle guiding the preparation of building foundation relates to the stability of the superstructure as influenced by the foundation. The modern concept of structural design of the footing of a building or other man-made structures originated from the recognition and identification of the behaviour of soils in natural condition and when subjected to load from the superstructure. In civil engineering practice of the modern time, the design is carried out in two phases — one for the natural foundation involving the soil aspects and the other for the structural footing, which forms an integral part of the man-made structural system and is involved in the process of load transmission on to the soil. Since soils in natural condition vary widely, both horizontally and vertically, even within the confines of the same structure, the soil design primarily is to arrive at the plan dimensions of the foundation, which would satisfy the requirements of bearing capacity and settlement of the soil under transmitted load. The depth of excavation occupies a predominant role in the design.

Soils originate by the weathering process of parent rock bodies, and are deposited by natural agencies of water, air, glacier, etc., and are subjected to subsequent biological activities. The variation of soil properties in horizontal and vertical directions are brought about by the intensity and extent of natural processes which are responsible for their origin and changes. The details of the formation of soil, changes brought about in their physical and chemical properties by various processes and agencies, etc., are not discussed here, being beyond the scope of the present work. The shallow foundations, therefore, should be placed at some identified depth below the land surface such as to avoid the zone of superficial soil which is susceptible to volume change or dimensional change arising out of various weathering and near surface biological and bacterial effects. The foundation prepared with slabs of stones in contact with soil below is meant to receive load from the superstructure and transmit it to the soil below. The cardinal principle involved in the foundation is that in the above process of load transmission, no component of either the superstructure or the foundation should experience distress of any kind.

This subject today has developed into a very sophisticated branch of civil engineering known as 'soil mechanics'.

The discussions at length on this aspect have been prompted by the processes described in *sūtra* 2.5.2 and their startling similarity with the present-day practices in civil engineering. These are (i) excavation of soil mass, (ii) dewatering and making them moisture free, and (iii) use of strong and durable stone slabs (concrete or bricks in present practice) over the bottom of excavation.

The earliest shallow foundation were strip footings in masonry — rubble or brick, and eventually in concrete. These foundations had to be substantially thick even when the flexure on account of the transmitted loads was of small magnitude.

The aim of soil design is to arrive at the plan dimension, which would satisfy two requirements from the soil side, viz., (i) bearing capacity, and (ii) settlement. The 'capacity' in 'bearing capacity' indicates that it is an ultimate concept and hence it is not a 'safe' or 'working' or 'allowable' quantity. Bearing capacity refers to the ultimate, that is the maximum load the soil can bear or sustain under given conditions. It, therefore, involves the application of loads to the soil for which a medium such as a footing is needed. In other words, so far as the bearing capacity of the soil is concerned, the footing just plays the role of a medium to load the soil with (Kurian, 1992).

Practical application of this principle has been obtained from archaeological excavations carried out at Kumrahār (Pāṭaliputra). A pillared hall of the Mauryan period, revealed the nature of foundation [*fig*. 1(a) and 1(b)]. Incidentally, this pillared hall,

fig. 1 (A) : Timber platform in the foundation, Pāṭaliputra excavation.
(Courtesy : Archaeological Survey of India)

fig. 1 (B) : Foundation Pillars, Pāṭaliputra excavation.
(Courtesy : Archaeological Survey of India)

according to archaeologists, is perhaps the earliest building in south Asia, employing monolithic stone pillars. The

> pillared hall was constructed by digging a series of 80 foundation pits in the alluvial silt, each between 1.52 to 1.83 m in depth and about five feet square. In each pit a 15 cm layer of rammed blue clay was laid and on this a timber platform of 1.4 m was placed, consisting of at least two layers, at right angles to each other. The pillars were erected on these platforms. Over this foundation a plinth, about 1.07 m in height, was raised and a further section of the shafts of the pillars was therefore buried. The excavators estimated that a complete pillar was probably 9.6 m in length, but that about 2.7 m of this total was buried beneath the floor of the raised plinth. To the south east of this hall, a canal of 3 m depth and 12.6 m in width had been excavated. The excavators conjecture that the massive timber platform immediately south of the hall formed the foundation for a broad ramp or stairway leading up from the ground level to the floor.
>
> — Allchin, 1995, pp. 237-8

The conclusions by the excavators are logical. Till about the advent of reinforced concrete, only by about the second half of the nineteenth century, the practice of foundation laid on stones, bricks and timbers was in vogue. It appears that some of the famous buildings in India constructed even in the earlier part of this century, have timber cribs/grillage as the sub-structural foundation. The load transmitted from the superstructure on to the footing results in a reaction in the bearing soil which bends the footing. The pressure stresses (reaction) in the bearing soil will depend upon the rigidity of the footing and the physico-mechanical properties of the soil. The purpose of the footing, besides this, is to evenly distribute the stress on to the footing. The underlying basis of this theory appears to have been learnt at the time of Kauṭilya through observation and practice. The logic of taking the foundation below the ground surface has already been explained. And in conformity with the theory, the ancient buildings, as the pillared hall of Pāṭaliputra, indicate the practice of providing the necessary precautionary measure of extending the foundation of a building to required depth below ground surface.

While on this issue, that is consideration of the properties and behaviour of superficial soil deposits, as related to the foundation of buildings, pillars, etc., another aspect attracts attention. This is concerning the stability of the natural soil in an excavation, as in case of the moats. *Sūtra* 4, chapter 3 of Book 2 describes the construction of ditches around forts, as protective measure. It states that

> . . . three moats to be dug round it, at a distance of one *daṇḍa* from each other, fourteen, twelve and ten *daṇḍa*s broad, three quarters or half of the breadth deep, one-third (of the surface breadth) at the bottom or square with the bottom, paved with stones or with the sides (only) built of stones or bricks, reaching down to (natural springs of) water or filled with water coming from elsewhere, with (arrangements for) draining excess water, ...

Points requiring attention are : (i) dimensions of the moats (i.e., excavations within ground) — which are stipulated to be (a) 84, 72 and 60 feet broad (one *daṇḍa* is equivalent to 6 feet) at the surface; (b) 42, 36 and 30 feet deep, being half of the respective widths; or (c) 63, 54 and 45 feet deep, being three quarters of the surface width. The bottom width could correspondingly be 28, 24 and 20 feet, taking one-third of the surface width as stipulated; (d) the moat can be square with the bottom; (ii) construction procedures — whether in soil or in rock.

In the stipulations, the bottom width has been kept as one-third of or equal to the surface width. In the latter case, the shape of the excavation becomes a square. The depth may vary either half or three quarters of the surface width. The descriptive dimensions of the moats are representable in geometrical figures (see Appendix 1). Simple trigonometrical calculations with widths and depths bring out the angles of slopes of excavation. In case of the stipulation in (d), when the shape of the excavation is square, that is when surface and bottom widths are equal, the side slopes obviously are vertical, or slope angle is 90° irrespective of the depth. In other cases, that is when the depth varies from half of the surface width or three-fourth of the same, the slope angles vary from 56° to 66° with the horizontal.

Thus, from the dimensional proportions stipulated in case of artificial excavations, slope angles of the excavations can be derived from simple trigonometrical relations. This angle and the height, whether in natural slope or in artificial excavation play very significant role in the stability of excavated slopes. Their significance in the stability considerations was of quite recent discovery. The first major contribution towards the study of the stability of clay slopes was due to Collin, a French civil engineer who was confronted with the problems of bank slips while constructing canals, around 1846. He realised that the failure surface was curved and that soil moisture plays an important role. His work did not receive much recognition till about the earlier decades of the twentieth century, before the First World War. A number of incidents occurred in different countries. During the excavation of Panama Canal through the narrow corridor between North and South America, huge slides took place from time to time. Many failures, almost at the same time, took place in earth dams in United States of America. In Sweden, railway embankments and cuttings suffered from extensive slides. Similar difficulties of slope failure were faced while excavating Kiel canal in Germany. All these incidents led into the research work towards studying the soil mechanics as related to stability or failure of natural or artificially excavated slopes within soil and rock (Dept. of Scientific and Industrial Research Laboratory, 1959, p. 501). The various technical methods of determination of the physical properties of soils and methods of calculation of slope stability and safety factors, as related to these and the slope configurations, etc., are beyond the scope of the present work. But generalised conclusions and rule of thumb, taken as a guide in modern day practice are briefly reproduced in Appendix 1, Section A. Experience has shown that slopes of 1.5 (horizontal) to 1 (vertical), or slope of about 34° are commonly stable. As a matter of fact, the sides of most railway and highway cuts less than 6 m (20 ft) deep rise at that slope, as do the sides of many deeper perfectly stable cuts. Therefore, a slope of 1.5 to 1 can be considered the standard for highway and railway construction. The standard slopes for flooded cuts such as those for canals range between 2:1 and 3:1. Steeper than standard slopes should be established only on rock, on some residual soils, on dense sandy soil interspersed with boulders and on true loess (Terzaghi and Peck, 1967, p. 415[1]). Natural or cut slopes can be made stable by providing some kind of restraining meaures, like pitching the slopes with brick, stone or timber cribes.

In modern practice in soil mechanics, the angle of slopes, for a given factor of safety of an embankment or cutting, can be obtained by knowing the soil characteristics. While in modern science, these properties are determined by laboratory tests and analyses, and factor of safety calculated using these parameters, extensive observation may, on the other hand, provide the range of slope angles at which enbankments or cuttings remain stable in different soil formations. From that it is possible to derive the relative dimensions of excavations that would render stability to them. This is what is found in the *sūtras*, which specify the dimensions of excavations. The generalisation about dimensions of excavations must have been put forth only after learning the laws and relations of the related dimensions by observations — a true expression of inductive logic — in the sense that they were derived by a certain inductive reasoning from the facts of experience. The process is one of the accepted sources of knowledge, where a general proposition is constructed from given results and consequences.

The instructions pertaining to the construction of earthen ramparts reveal yet another very significant aspect in soil mechanics. It has been stipulated that the rampart should be 6 *daṇḍa* (i.e., 36 ft) high at a distance of 4 *daṇḍa*s (24 ft) away from the edge of the moat. The foundation of the rampart should be so strongly bonded with the underlying formation that the earth with which the rampart is

---

1. Slopes expressed in gradients, i.e., in relation to vertical and horizontal lengths, can be expressed in angles as below:

Vertical	Horizontal	Angle
1	1.5	33.69°
1	2	26.56°
1	3	18.43°

constructed does not yield, flow or spread out. In order to prevent such an eventuality, it has been stipulated that the earth should be 'pounded by elephants and bullocks (and) having (on the sides) clusters of thorny bushes and poisonous creepers' (*sūtra* 5, section 21, Book 2). It cannot escape notice that this process involves compaction of soil and protecting the exposed surface of the constructed earthen structure from shear feailure and also erosion, by planting suitable plants. Technically speaking, consolidation of loose soil is brought about by decreasing the water content of a saturated soil, without replacement of the water by air. This is achieved in the modern practice of construction of earthen dams, earthen embankments, etc., by repeated application of pressure on the soil by means of various types of mechanical rollers. One such found to be very effective and hence widely used in the construction of earth dams and similar structures is the sheeps-foot roller. This consists of a hollow cylindrical steel drum on which projecting metal feet are mounted. The principal characteristics of sheeps-foot rollers in compacting soil are the foot pressure and the coverage of ground obtained per pass. Excavated earth from borrow pits with which such sructures are constructed are normally in chunks. An indivdual chunk of soil cannot be compacted by any of the methods of applying pressure of short duration or vibration, since neither of these processes produce more than an insignificant change in water content of soil. However, sheeps-foot rollers are effective in reducing the size of the open spaces between the chunks, thereby aiding consolidation of the soil. Similar effects were sought when loose earth was to be pounded by elephant or bullock, to make the earth perfectly compact.

Prescription of planting thorny bushes and poisonous creepers may apparently appear to be an attempt to stave off animals from grazing and destroying the plantation on ramparts and thereby loosening the earthen structure (*fig.* 2). But another obvious fact implict in the instruction is to cover the earthen surface with plants, which help in binding loose soil with their roots. Poisonous or throny bushes, by keeping off animals, would provide protection against disturbance of the earthen surface, which would otherwise subject it to loosening and thereby making it easy target for erosive action of

Fig. 2: Artists impressions of Kauṭīlyan Fort

wind or rain. It is a common practice to resort to turfing of earthen slopes with grass and other suitable vegetation, vegetative cover serving as a rip-rap surface.

> Vegetation of suitable types has been successfully employed in stabilising slopes. Apart from reducing the moisture content of the soil (and hence increasing the shear strength), the network of roots acts as a reinforcement in the soil. . . . Vegetation can also act as a blanket on the surface layers of soil by limiting the softening effects of water and reducing erosion. — D.S.I.R, 1959, p. 509

These processes and measures, so transparently described in *Arthaśāstra*, are in vogue in the modern practice, following the recent development and progress of soil mechanics as a branch of modern science and technology.

Buildings and towers

Several *sūtras*, viz., 2.3.7, 2.3.10, 2.3.11, 2.3.16, 2.3.18, 2.3.21, 2.3.22 specify dimensions of building, their component elements — chambers, halls, turrets, parapets, pillars, etc. Dimensions of engineering structures are not arbitrarily assigned, but depends on the knowledge of the properties of materials with which the structures are constructed, like their strength and capacity to withstand forces acting on them without failure, the pattern and magnitude of load acting on them, etc. Most practical structures consist of complex arrangements of many components. Those are determined on the basis of analysis of forces acting on them. And the analyses are complicated further because the forces experienced by a structure are variable and sometimes unpredictable. A building experiences various forces, like the load of materials with which the building is constructed, wind forces lashing it, the load of moving objects, ground movement due to vibration, say by earthquakes, and so on. The distress caused by such forces, the way such forces are transmitted along the structural components, by way of compression, tension, shear or gyration, etc., and the response of materials to such forces are the basic information on which the components are designed. Dimensions of each component are related to such analysis. Modern civil engineering is primarily based on the knowledge of forces and the behaviour of construction materials and foundations. Intricate design of the modern structures has been possible with the advancement of these knowledge including the same on the properties of construction materials. Till then practical experience played the predominant role in adopting technology. And this bears the sign of repeated experimentation by actual practice. For instance, practice of agriculture since the dawn of human civilisation demonstrates repetitiveness of a process with great predictability, learnt by experience through generations. The age-old practice of agriculture has become understandable today as a science with the

> rise of cybernatics or systems theory — the theory of how the diverse elements of systems interact through time to produce change. Even in oral traditions, agriculture has always been perceived as a system; hundreds of generations of farmers have understood the interaction of stock, soil, water supply, climate. Today these elements are understood better than ever; they can be quantified; they can even be controlled. — Mayer and Mayer, 1974, pp. 83-4

In other words, the codes of practice, formulated in words, unlike those in the normative pattern of modern science, must have been done after repeated practice and testing through reproducibility of similar results from similar process of practice.

> . . . a correct idea can be arrived at only after many repetitions of the process leading from matter to consciousness and then back to matter, that is leading from practice to knowledge and then back to practice.
>
> — Mao, 1963, p. 727

This form of knowledge is widely observable in *Arthaśāstra*. In the cases of construction of buildings, excavation in natural soil, process of stabilisation of soil against failure discussed earlier, significant meaning could be extracted out of such codes confirmatory to the modern theories. The dimensional aspects of buildings and other engineering structures codified in the *sūtras* referred to earlier prove that they must have been derived out of the basic principles underlying the design of structures, that is on the knowledge of interaction of forces on materials and the response of materials to such forces. The *sūtra* on the specification of the dimension of pillars (2.3.18), has been analysed in details in Appendix 1, Section B.

The other stipulation in the same *sūtra* relates to the depth of foundation — which should be twice the width. The argument in favour of this postulation, relevant to the stability of a structure, has already been discussed earlier in relation to the preparation of foundation of a building.

The brief description of the various aspects of construction codified in *Arthaśāstra* demonstrates a full gamut of principles underlying the civil engineering of today and these may be summarised as under:

(i) Physical nature of ground (topographic characteristic) where a building should be constructed.

(ii) The properties of soil on which a building is placed, nature and dimension of excavations, measure to be taken against potential failure of slope or excavation, etc. — revealing the basic perceptions of the science of soil mechanics.

(iii) Dimensions of the building components, and their relative proportions, obviously derived from the experience of their stability, when assessed in terms of the interaction of forces and materials (of construction).

(iv) Foundation preparations and depth of foundation.

(v) Materials used for construction, etc.

Descriptive account of specific locations of different kinds of buildings in a town or other habitations, of the king and of other professionals, their spatial distribution within the city, within fortified towns and also in villages, types and structures of building, etc., occupy prominent place in *Arthaśāstra*. The layout and design of such buildings are found in case of (a) royal residence (Book 1, chapter 2), (b) buildings and forts (Book 2, chapter 3) and (c) buildings related to public utilities (Book 2 chapter 5).

ROYAL RESIDENCE

The royal buildings shall be erected, facing north or east, on a good building site, in the middle of the residential areas of the four *varṇas*, (2.4.6 and 2.4.7) and as per the recommendations of experts in the science of building (2.3.3). The points to be taken into account in selection of appropriate site for settlement are that it should have advantages of natural or artificial protective measures and should be near sources of water. These have been enumerated in Book 2, chapter 3 in connection with construction of forts. Obviously, the safety, security and availability of provisions and other necessities of life were to be considered in selecting the appropriateness of settlements with royal residence within. The royal residence should be on the square immediately north of the city centre (2.4.7). The royal building should have moats and ramparts all around, with suitable gates, as protective measure and the palace should have a number of halls. The living chamber should be at the centre, and should be constructed in a manner similar to the construction of store house (2.5.2-3). This instruction appears to be related to the preparation of foundation enumerated for the cellar/store house. The chamber should be provided with concealed passages within walls. Other form of construction, considered for the security of the king, may be an underground chamber with its opening covered by the wooden image of a deity and having anastomosing subterranean passages. The palace may be located above it with concealed passages for entry and exit provided with mechanism, which allowed instantaneous entry into the subterranean chamber.

> The floor of the *prāsāda* is so constructed that with the release of a mechanism it can fall down into the cellar below.
>
> — *Sūtra* 1.20.2; Kangle, vol. 2, footnote in p. 49 (please see Appendix 1, Section C).

All these measures are to be provided to tackle all kinds of conceivable emergencies, including calamities. At the backyard should be rooms to be used by ladies during confinement, as a sick room, and also for trees and water. The latter prescription does not really fit well if taken literally. It may be assumed that the sick room is to have provisions for enough water required for treatment, ablutions and that there should be appropriate plantations, as decorative and soothing greenery which provides recuperative beauteous atmosphere for the sick. In the background of very sensitive and keen interest displayed elsewhere in this treatise about natural environment and cultural mindset, this explantaion about the association of floral plants may not appear to be baseless. The rooms for princes and princesses are to be located in the outer periphery of the residence. There should be dressing room, council chamber, assembly hall and a separate chamber for the prince's governor, all to be located with reference to the central hall. Dwelling places for servants and other helping hands in the royal residential service have been prescribed.

The layous of the royal palace with rooms, chambers, halls, etc., as necessary in the private life of the king, for discharging royal responsibilties, like councils with ministers, holding meetings, etc., display a very high order of architectural concept. Details of space provision for stores, sick room, education and training of princes by residential governors, soldiers and guards to keep vigilance, etc., have been enunciated. Intricate constructions of structures, like underground chamber, labyrinths concealed within walls, other camouflaging mechanisms for hiding and speedy exit, have been specified with details (please see Appendix 1, section D). Presence of such structures, as revealed by hitherto available archaeological findings, endorses that these were not merely imagination, but at least some constructions used to be practised in reality. Architectural layout and designs of royal palaces of much later centuries proved the viability and practicability of these concepts.

CONSTRUCTION AND CONSTRUCTION MATERIALS

Besides the important aspect in connection with the foundation preparation, the *sūtras* prescribe also the construction materials. Timber, bricks and stones as construction materials have been mentioned. Presence of timber and bricks have been found in profusion within the excavations carried out of sites proved to be of Mauryan period. Stones are not present in comparable proportion. This fact has been taken by some authority to relegate the date of *Arthaśāstra* much later than fourth century BC, on the ground that the use of stones in building construction entered into Indian practice much later. As against such a postulation, the simple explanation may be offered. Since stones are not readily available at manageable distances within Gangetic valley, stones did not find the place in the civil construction here, and all the ingenious methods were mobilised to make use of the available resources. Even then, it has been proved that the use of stone was not unknown to the then builders, at least wherever they were available [*fig*. 3 (a) and (b)].

> The discovery of the remains of a pillared hall, resting upon a substructure timber, and consisting of pillars of Chunar sandstone, bearing a distinctive polish, was made by Spooner, an officer of the Archaeological department, in 1913. . . . As yet no comparable stone structures associated with the Mauryan period are known from Pataliputra, but in the absence of systematic search this is scarcely unexpected. From the other excavations conducted there from time to time, and from the chance finds of public works in various parts of the city, a number of striking and suggestive fragments of polished stone sculptures and architectural elements have been discovered. . . .
>
> — Allchin, 1995, p. 204

The use of timber in building construction, including in the foundations of building may not only be considered on its availability, but some of its inherent properties might have been known and hence used. Absence of stones near about, as reliable strong material, known from the dawn of the human evolutionary history, forced the people to explore these properties in naturally available other types of materials, which could suitably be used in buildings. Timber is one such, and it is quite strong in tension, which means that a timber beam can withstand a considerable vertical load without fracture. That is the reason why the floors of some

fig. 3 (A) : Stone rampart around Rājgīr. (Courtesy : ASI)

fig. 3 (B) : Excavated brick structure, Kumrahār. (Courtesy : Archaeological Survey of India)

modern building are still made of timber besides aesthetic beauty they offer. It may not be without any logic if it is concluded that this properties of the timber were well known, and their use as grillage platform, as described earlier, in the foundation, as found beneath the pillared hall at Pāṭaliputra, must have been done consciously.

There is a very interesting indication about another type of construction material and process in *sūtra* 4, Book 1, chapter 20, as a measure of protection against fire. Other processes in the same *sūtra* being shrouded in rituals appearing as quaint occult practices, this interesting part remains rather veiled behind the verbose descriptions of the rituals.

It is said that when clay is mixed with ashes produced by lightening (*vaidyutena' bhasmana*) and water derived from hails (*karaka-vāri*) and applied to the walls, it becomes fire-proof. Like other parts of the same *sūtra*, apparently this process also appears to be odd and illusory. When almost similar process described in *Matsya Purāṇa* (about fourth century AD) is scanned, it becomes clear that the statement contains some sense and meaning to be reckoned with. It states

> *samudra saindhavayava vidyudagdha ca mr̥ttikā tayānuliptam yadesom nagina dahyate nr̥pa*

> It may be seen that *vaidyutena bhasmana* has been expressed as *vidhyudagdha ca mr̥ttikā* meaning 'earth burnt by lightening'. — Bhattacharya T.P., 1986, pp. 84 and 127

Vaidyutena, though normally means 'of or proceeding from lightening' may actually mean brilliant fire, and *bhasma* meaning 'ash' produced by such a fire. Such a situation may be conceived of ash derived during volcanic eruptions, a meaning, which fits well with the product and the process described. But geological history of India does not endorse this natural process to have taken place after the advent of man. The process, therefore, has to be sought in comparable physical one. Like the porous volcanic *lāvā* (pumice), produced due to escape of gas, over-burnt bricks also produce pumice like structure. Mixture of volcanic ash or pumice dust with lime produces a light variety of cement, known as puozzolana, which was used in building constructions by Romans since very early period of history. In India, brick-dust or brick-pumice dust mixed with lime is used as cement mortar since early times and is still in use. Thus, *vaidyutena bhasma* can be regarded as pumice (brick) produced by excessive fire heating. This material is of importance in the production of lime-mortar by its appearance only, but over-burnt brick contains a concentration of such elements as iron and alumina, which in combination with lime yields the cementing property. The meaning of *karaka-vāri* is not properly understood, neither has it been explained anywhere. *Karaka* means the 'hail' or even a 'water-vessel', and 'coconut shell' (Monier Monier-Williams, p. 254). Water derived from hails may connote very cold water. Nearest to the word *karaka*, *karkora* means the lime-nodules or *kaṅkara* as called in Indian parlance (Monier Monier-Williams, p. 250). But replacement of *karaka* by *karkora* may not be permissible unless it is proved that in the original manuscript, there is scope for such an insertion due to indistinct handwriting or missing letter or the like. That such possibility cannot be ruled out altogether, is proved by the fact that in Devanāgarī manuscript the word *karakabarina* is omitted and in the *grantha* manuscript the word *kanaka* is used instead of *karaka*, (Kangle, pt. 1, p. 25). Thus, insertion of *karkora* in place of *karaka* would have given an appropriate process of producing Puozzolanic cement by mixing pumice-dust with lime. In its present form, as stated in the *sūtra*, the process may be conceived of as of kneading the clay with pumice-dust and cold water. The mixture would offer stronger bond to the clay, and walls plastered with such material would certainly be more resistant to fire, compared to timber structures, which were in vogue in those days. Nearest to this, another word *kalka* is to be found in *Bṛhat-Saṁhitā*, which means a paste used as plaster or cement, also called *yoga* (mixture) (Acharya, 1927, p. 124). *Kalko'yam vajra lepakṣaḥ* in *Bṛhat-Saṁhitā* indicates to adamantine glue (*vajralepa*) prepared with various organic and inorganic materials. This was in use in the construction of temples, mansions, etc., because that was very essential for fixing idols in construction of walls and the like (Bhat, 1992, p. 546). In this regard observation has been made that

> Artisanal chemistry, which was very well developed, not only permitted numerous alloys preparations, but was also very advanced in the treatment or use of various minerals, particularly mercury salts, together with organic, vegetables, and animal products. All sorts of preparations resulted, from the most varied drugs to the stucco and cements used in construction and the production of various objects. Included in the specialities of Indian technology are several very hard cements (*vajralepa*, 'adamantine paste') obtained with lime and resins first extracted and blended for decoration, then reduced to liquid form by prolonged periods of distillation. . . .
>
> — Filliozat, 1969, p. 329

The architectural concept in consonance to the physical characteristic of the ground used for placement of buildings and spatial distribution of residences to be occupied by people of various professions, is well articulated in the chapter dealing with the layout of the fortified city (Book 2 chapter 4). The royal residence should be at the northern part of the city centre, houses of preceptors and chaplain, the places for sacrifices and for water as well as councillors should occupy the north-east part; the stables and magazines in the south-east part; the eastern quarter should be occupied by the dealers of perfumes, flowers, etc., along with kṣatriyas; store house, accounts and account office, and workmen's quarters should be located in the south-east part, store house for forest produce in the south-west part and grain dealers, factory and army officers, dealers of cooked food, wine, meat, courtesans, dancers should occupy the southern part of the city. In this manner, spatial locations for people engaged in various activities have been specified within and outside the enclosures of the fortified city.

Moreover, each group of people should have fixed boundaries of their household according to the necessity for respective activities. Within the boundaries, flower and fruit gardens are to be cultivated, and should have space for storing provisions. A cluster of ten families should have an enclosed space for well for water supply. Sanitary measures, space between houses, provisions for lighting each household, protection of exposed part of the house from damage from rains, etc., have been incorporated in the process of house-construction (Book 3, chapter 9).

In the different codifications of buildings, their specific locations with respect to each other and within the defined boundaries of a fortified city, the basic philosophies of architecture have implicitly been embodied in them. It is accepted that

> In the transformation of the environment, architecture has a peculiar part to play. This arises not merely because buildings constitute such a large part of man's daily surroundings; but because architecture respects and focuses such a wide variety of social facts; the character and resources of the natural environment, the state of the industrial arts and the empirical tradition and experimental knowledge that go into their application, the process of social organisation and association, and the beliefs and world-outlooks of a whole society. . . .

Moreover,

> The architect confronts human needs and desires with the obdurate facts of site, materials, space, costs; in turn, he moulds the environment closer to human dream. And in a social sense, architecture is more advanced than any purely mechanical technique because good building has always embodied, as an essential element in both design and operation, the understanding and expression of organic human purposes. . . .
>
> — Mumford, *op. cit.*, p. 403

Architectural specifications of different types of buildings, be these meant for fortification, watch towers, spaces to resist outside attacks, or attacking enemies, service roads for movement of watchmen and soldiers, royal rooms within forts, ramparts, parapets, exit doors, escape passages, etc., or buildings, ware houses, storage structures for grains, treasure house and others reveal that the perception of living standards, amenities and other functional components of the buildings had been of highest order, comparable to the present-day concepts and practices.

In case of community dwelling places (Book 3 chapter 8), the *sūtras* demonstrate the elements of effective neighbourhood planning — another modern concept in town planning and building-architecture.

> Effective neighbourhood planning, through imaginative and carefully considered use of outdoor space, by introducing trees, and by judiciously arranging the housing units into pleasant groupings, can achieve the amenities of spaciousness, light and air, and visual satisfaction, even in neighbourhoods of the humblest and smallest dwellings. For reasons of both economy and social desirability, the basic unit upon which the neighbourhood is planned is a single storey house, with independent entrance and enclosed private yard.

The building laws propounded in *Arthaśāstra* bear a very glaring similarity with the modern municipality rules. It has been observed that

> the laws of the *Arthaśāstra* on housing questions are surprisingly progressive and enlightened for Kautilya's age. They correspond to the modern municipal building laws and are vaguely in the nature of house planning. But it must be remembered that whereas the modern house planning is supposed to aim at the comfort and convenience of the members of the community, the primary object of the Kautilyan code is to effect a reduction in or removal of the causes that contribute to discord and friction amongst neighbours.
>
> — Sharma, 1950, pp. 3-10

To wrap up discussion on architecture, town planning and civil engineering constructions, contained in *Arthaśāstra*, and comparing these with the Western practices, the Indianness and 'vernacular architecture' are still discernible. The structure, types of buildings, town planning, etc., resources available and utilised and conformity to the physical environment truly define the vernacular science that was indigenous to the country. Archaeological findings, so far obtained, confirm this, though these demand much more extensive and intensive analysis by competent experts.

The preceding discussions on the establishment of habitations, their groupings, construction of fortified cities, forts, layout of township, royal residence, residence of common people, positioning of reserve forests and forests of other kinds, sanctuaries, etc., focus on two concepts, viz., orderliness and geometric attribution to space — two formidable innovative intellectual framework of the time.

Thinking of those, it transpires that the concepts are actually not unconnected to the historical milieu of that time. The time to which Kauṭilya belonged may well be identified as the second phase of urbanisation in Indian history,

> the first having been that of the third millennium in the Indus valley. The antecedents of this second urbanisation point to a shift in geographical location from the nuclear area of the third millennium. The Indus civilisation had declined by the middle of the second millennium BC, seemingly unconnected with the earlier copper-age civilisation. Technologically the new urbanisation was based on iron, the widespread domesti-cation of the horse, the extension of plough agriculture, and a far more sophisticated market economy than that of the earlier period.
>
> — Thapar, 1975, pp. 119-20

The seed of the development was immanent in the then existing social and political situations. The country, particularly south Bihar, which included the Magadha territory, and was considered to be the core of Āryavarta in the Jain and Buddhist literatures (Thapar, p. 131), was suffering from the aftermath of a massive foreign invasion, and in Magadha a dynasty was just thrown off. A new order was to be establsihed in a situation which, by all logic and reality, can be considered as comprised of not only political but social anarchy and chaos. The approach to establishing the authority called for introducing discipline in different spheres of society, consolidating means of production and power; defining the natural resources and their appropriate utilisation; collection of revenue, and the like — in short ushering in of orders in every sphere which would ensure the stability of monarchial system. Spatial and temporal experiences were the prime constituents which were mobilised in evolving the measures for centralising the power, as evidenced from the creation of centres of administration, in the form if *sthānīyas*, *droṇamukhas*, *karvaṭika*, *saṁgrahaṇa*; promulgations of rules on the authority over natural resources, including land, minerals, mines, water; right over the revenue on business and trade; consolidation of the source of wealth, by establishing ware houses, underground cellars; raising army, establishing armoury, prison, etc., being sources of power; imposition of punitive measures for violation of regulations, adopting and practising neighbourly attitudes and behaviours, etc.

Monarchial security, bureaucratic patterns introduced in the system and political and social imperatives dictating strict adherence to orderliness must have shaped the mindset to precision attitudes to everything, including space. In the layout of towns, roads, forts, residences, even in the arrangements of rooms within a house, etc., the space was 'geometrised' and aptly be said that 'space became specialised room by room' (Mumford, p. 115). Besides the economy of space, the geometrical concept of space manifested the sense of aesthetic beauty in its appropriate allocation and use. Thus, in the concept of orderliness, imparting material attribute to space in confining it by geometrical shapes, trimming and beautifying by pleasing layout and optimum utilisation, we find not only the signs of breakthrough in the concept, but transcendental philosophy which moulded the architectural technology on a higher plane, comparable to what it assumed later on till our time.

4

Land, Agriculture and Irrigation

ARTHAŚĀSTRA prescribes various attributes of a settlement/country, which indicate its excellence and richness. As regards land, it has been indicated that such country should be

> devoid of mud, stones, salty ground, uneven land, . . . endowed with agricultural land . . . with protected pastures not depending on rain for water. . . .
>
> — 6.1.8

Land and agriculture occupied prime positions in assessing the wealth of the country,

> Agriculture, cattle rearing and trade — these constitute economics . . . as they yield grains, cattle, money, forest produce and labour.
>
> — 1.4.1

All forms of land use — in the most profitable way, both from economic and environmental considerations, are to be found in the way these are desired to be distributed, specified for the use and promulgation of punitive measures for violation of these. Elaborate instructions are given on census procedures and land survey leading to the classification and categorisation of the quality of land. These were made mandatory task of the Administrators. The tasks of the Administrators were defined clearly.

> Dividing the countryside into four divisions, the Administrator should cause to be entered in a register the number of villages, classifying them as best, middling and

lowest, (recording) this is exempt from taxes, this provides soldiers, this much is (the revenue in) grains, cattle, cash, forest produce, labour and produce in place of tax.

— 2.35.1

Further

He should record the number of villages by fixing their boundaries, the number of fields by an enumeration of ploughed and unploughed (fields), dry and wet fields, parks, vegetable gardens (flower and fruit), enclosures, forests, structures, sanctuaries, temples, water-works, cremation grounds, rest houses, sheds for drinking water, holy places, pasture lands and roads, (and) in conformity with that he should keep records of the size of boundaries, forests and roads, and of grants, sales, favours and exemptions, concerning village boundaries, and fields, and (keep records) of houses by an enumeration of taxpayers and non-taxpayers. — 2.35.3

There would be one Superintendent of Pasture lands,

And he should establish pasture land in regions between villages. — 2.34.6

He should clear low lands and forests of the danger of robbers and wild animals. — 2.34.7

Moreover

In waterless regions, he should establish wells, water-works and springs, also flower and fruit enclosures. — 2.34.8

The duties of officers are available from descriptions of Megasthenes as,

Of the great officers of the state, some have charge of the market, others of the city, other of the soldiers. Some superintend the rivers, measure the land, as is done in Egypt, and inspect the sluices by which water is let out from the main canals into their branches, so that everyone may have an equal supply of it. . . . They collect the taxes, and superintend the occupation connected with land, as those of the woodcutters, the carpenters, the blacksmiths, and the miners. . . . —McCrindle, *op. cit.*, pp. 86-9

Land distribution

On completion of census, the modes of distribution of land, which belonged to the king, for their potential use, tax collection, benevolent puirposes, etc., have been formulated. And much emphasis was given on the manner of use of the land to be deployed for cultivation, horticulture and forestry. In order to place the land for best and appropriate use, the recipients were to adhere strictly to the contractual terms and provisions. Any violation was to be met with penalty in the form of fines or even dispossession of land. Royal assistance and support were guaranteed to those who carried out proper cultivation, taking care of sustaining the quality of the land. These are enumerated in various *sūtra*s as follows:

- Land to be granted with exemptions of tax and fines with inheritance right but without right of sale or mortgage, to those practising religious acts — priests, chaplains, brāhmaṇas, physicians, heads of departments, *gopa*s, *sthānika*s, elephant trainers, etc. (2.1.7).
- The king should allot to taxpayers arable fields for life (2.1.8).
- Those who make unarable land arable should have right to it (2.1.9).
- Exemptions from tax, etc., should be granted at the time of settlement and in doing so, the king should act benevolently showing due considerations (2.1.17 and 2.1.18).
- The tillers of land should only be favoured and those who keep the land fallow (2.1.10 and 2.1.12) or do not sow in time (3.10.8 and 3.10.14) should loose the possession or pay fines or make good the loss to the treasury. The land can be taken away for awarding it to be cultivated by village servants or traders (2.1.11) or some other may use it for five

years and return it after receiving compensation for his exertions (3.10.14).

- Non-agricultural land may be allotted as pastures for cattle (2.2.1) in addition to establishing pastures land in regions between villages (2.34.6).
- Such land may be utilised to create an animal park for Royal use, protected by moat, containing shrubs, fruit bearing trees, thorn-less trees, tamed animals, with water pools, etc., (2.2.3) or raise forests (2.2.5) for indicated forest produce (2.17.4 to 2.17.16) and also factories using forest produce as raw-materials. Forests attached to the produce forests may also be settled alongside (2.2.5).
- The cultivators were to be given due support in cash and kind, like grains, cattle and money (2.1.13), particularly when the land is cultivated by village servants and traders. These were to be repaid back at their convenience.
- The Director of Agriculture was also made responsible for carrying out cultivation through serfs, labours and bonded labourers, create seed-banks for various kinds of grains, flowers, fruits, etc. (2.24.1 and 2.24.2). He was supposed to stick to time and for that make timely arrangements of infrastructural supports of artisans, smiths, carpenters, plough (*karṣanayantra*), etc., (2.24.3).

The distribution of land appropriate for different uses in cultivation, forestry and pastures, with strict injunctions on their use, administrative control, supports in every sector and extension services to the cultivators indicate systematic and well organised agricultural practice prevalent at the time of Kauṭilya. That this was practised in the most scientific way is proved by the accounts of climatic influence and irrigation recorded in *Arthaśāstra*.

Agro-climatic concepts

Climate profoundly influences agriculture. Temperature, precipitation, wind speed and humidity are the most important factors controlling agriculture. Seasonality of crops conforms to these features and unseasonable changes may reduce productivity, damage or even destroy the entire crop. Certain crops are sensitive to certain times of the season, recording different weather condition. Many plants demand much water during the period of active vegetative growth, but they need a dry or almost rainless ripening period. In case of rice cultivation, it requires high temperature in order to germinate, bloom and mature. Some varieties of rice would not germinate at temperatures below 20°C (68°F). During the active growing temperatures must be much higher.

In general, moisture requirement for rice is about 1143 millimeter (mm) (45 inches), and the same for wheat vary between 254 and 635 mm (10 to 25 inches).

In regions of South East Asia with favourable temperatures, water requirement is often quite heavy. It demands annual rainfall of at least 1270 mm (50 inches), and 127 mm (5 inches) or more during each month of the growing season.

High summer temperatures and heavy monsoon rains, as in India, can be best utilised in the process of transplantation — that is seeding rice in nursery beds and transplanting it later on.

These are some of the knowledge of climates, which have been proved to be important features of agricultural practices, not only from observation, but calibrating these by the scientific methods of quantitative measurements. The climatic perturbations — leading to excess or deficiency of water, are amenable to remedial measures, like providing drainage to remove excess water, or providing irrigation to make up for deficiencies. These have been possible by advancement of science and technology. Man, unable to do much with climate in grand scale, has learnt to adjust by adapting to cropping pattern in consonance to seasonal variations.

It has been attempted in *Arthaśāstra* to delineate geographical zonations of the country based on the amount of annual precipitation (Map 2). Arid regions are those which receive annual rainfall of sixteen *droṇa*s, that is about 32 inches (about 810 mm), wet regions receiving annual rainfall of one and half times of that, that is about 48 inches (1220 mm). (About conversion of *droṇa* to modern measure

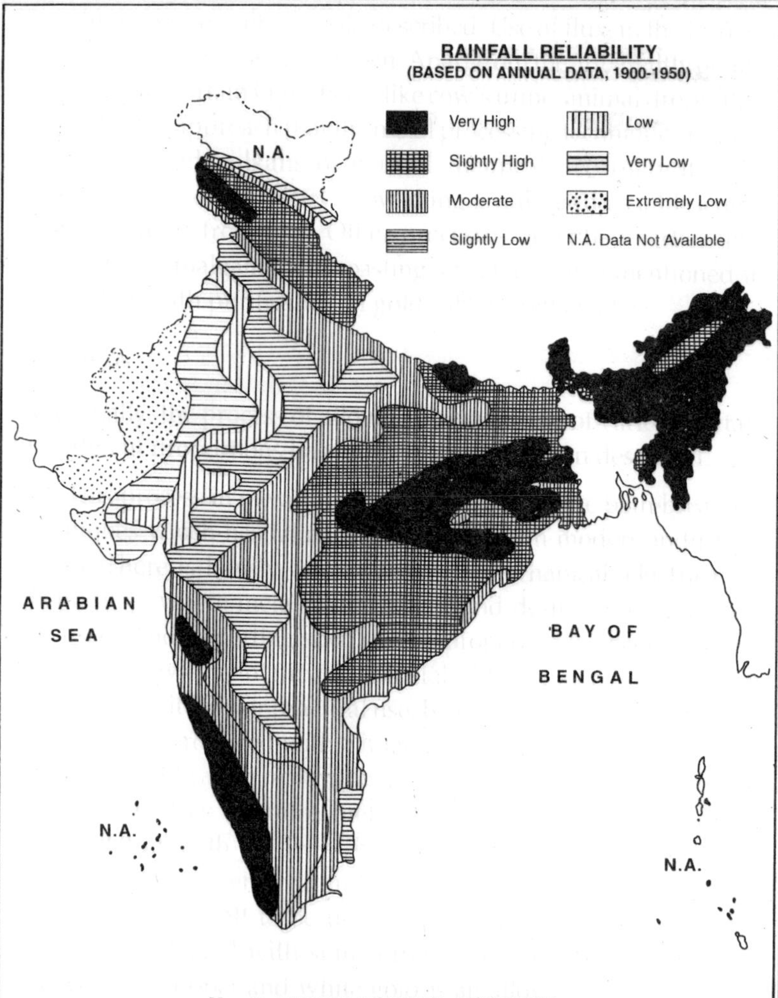

Map-2a RAINFALL ZONING COMPARED

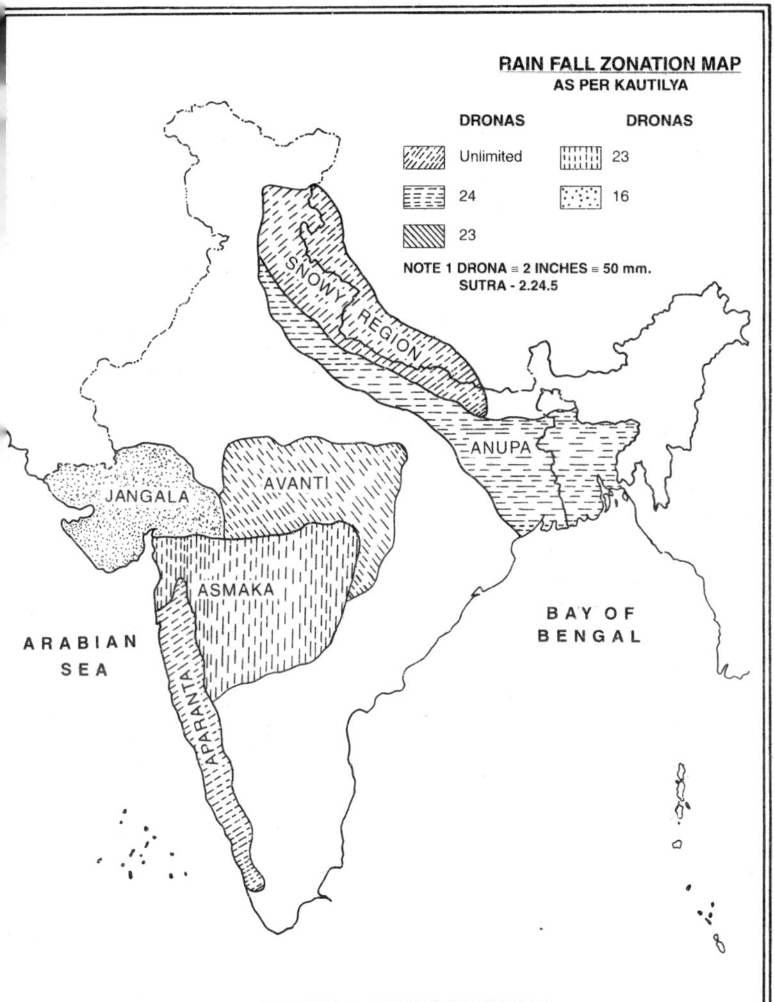

Map-2b RAINFALL ZONING COMPARED

in inch, footnote under 2.24.5 in Kangle, part II, p. 149 is referred. Conversion of inches to millimetre has been done as 1 inch = 25.4 mm.) Annual average rainfall in different locations has been indicated. The annual rainfall in the Mahārāṣṭra region (Asmaka), that is in the south-western region of India, is about 27 inches (685.8 mm), in Ujjain region (Avantī), that is on the northern slope of Satpurā range, it is abouat 46 inches (1168 mm), and in the extreme western border (Aparānta, which might be on the western slope of Sahyādri range, that is coastal Mahārāṣṭra, Karnataka and Kerala), and in the snowy regions (Haimanya, Himālayan region), the annual precipitation is unlimited (2.24.5). There used to be a system of measurement of rainfall, as is evident from *sūtra* 2.5.7 where it is stated that 'In the magazine, he should place a basin with a mouth one *aratni* (in width) as a rain-gauge'. (One *aratni* is about 18 inches diameter; Kangle, pt II, p. 149).

The time and amount of rainfall, which are beneficial to crops, have also been indicated. It is said that of the total rainfall, one-third amount in the first and last months of the rainy season, with two-third in the intervening months, are the ideal conditions (2.24.6). In *sūtra* 2.24.9 the intensities of rainfall have been indicated. The patterns of rainfall accompanied by winds and sunshine, supporting crop growth have been described in *sūtra*s 2.24.9 and 2.24.10. In addition to these observations, the method of forecasting rains by astrological observations of the constellation of Sun, Jupiter and Venus has been prescribed. On finding the time, probable amount and intensity of rainfall, sowing of different crops are to be taken up (2.24.11 and 2.24.15). Crops, requiring water in variable quantities have been indicated and their sowing time indicated during the segments of season (2.24.12 to 2.24.14).

The *sūtra*s referred express and depict the entire spectrum of agricultural meteorology-temperature, rainfall, wind, sunshine, cloudiness, etc. Modern-day development of agricultural engineering and meteorology is based on due recognition of these natural elements and the sophisticated gadgets only aid in their determination. It has been recognised and established that the essential requisities for successful production of crops are water, air sunlight and soil (this knowledge is not the first ever discovery of modern man, but perfection, to the extent possible, considering uncertainties and perturbations in nature, has been achieved in minor details with advancement of science and technology). Air temperature and soil temperature plays a vital role for the growth of crops. There are optimum temperatures for germination and process of crop growth. With increase in termperature, the crop-water use increases. Solar radiation is one of the most important factors in photosynthesis (ability, common in plants, to from organic compounds from water and carbon dioxide in the presence of sunlight, necessary for their survival and growth) and transpiration (physical process through which all plants lose a certain amount of water) of crops. Cloud amount is indirectly measured by observation of the cloudiness of sky. The international unit for estimating the cloud amount is the 'okta' or eighth of the sky. The assessment of the total amount of cloud, therefore, consists in estimating how much of the total apparent area of the sky, to the nearest eighth, is covered with cloud. For this purpose, one should observe the sky and imagine it to be divided into four quadrants. By imagining the clouds that are scattered in different parts of the sky to be put together, and how many quadrants the total cloud occupies should thus be estimated. The amount of cloud is 'zero' only if the sky is completely cloudless and 'eight' if the sky is completely overcast without gaps of any kind (NCERT, 1985). How similar the measurement of cloudiness of the sky expressed in *sūtra* 2.24.9! 'Three', 'eighty', and 'sixty' clouds can be taken to be expressing the 'okta', in different proportion of expression, comparable to what is practised today.

Climatic factors, which influence the crop germination, growth and final product, taken together, allow delineating geographical zonation, and similar attempt is evinced in *sūtra* 2.24.5. Similarly, rotational sowing of crops in different seasons has also been advocated (*sūtras* 2.24.10, 2.24.11 to 2.24.15). Recently, India has been divided in some 15 agro-climatic regions in order to suggest measures to overcome the problems of each region in the process of improving productivity. The demarcation of these agro-climatic regions has been based on homogeneity in agro-characteristics, such as rainfall, temperature, soil, topography, cropping and farming systems, water resources, etc. These zones are the Himālayas, Gangetic plains, plateau and hill regions of central, eastern, south-eastern and western India, coastal plains and hills, western arid region and island. The natural characteristics of these regions are well recognisable and they have distinctive agro-climatic set-up.

Crop water relationships have remarkably been dealt with. It has been stated that crops should be selected according to the availability of water in different seasons (2.24.19) and moisture content of different land (2.24.23). It is indicated that Śāli crops are best, vegetables are middling and sugar-cane worst, requiring more expenditure and being fraught with many uncertainties (2.24.20-1). The rhythm of agricultural practice has been expressed through the seasonality of crop cultivation. In addition to the climatic influence on such crops, a clear picture has been given of the mixed cultivation through the succession of sowing — to begin with, with rice cultivation, followed by lentils, barley, wheat, etc. (*sūtras* 2.24.12 to 2.24.15). Succession of crops from season to season has been referred. These are like: *vārṣika śasya* (rainy crops), followed by *haimana muṣṭi* (autumnal harvest in *Mārgaśīrṣa* (November-December); *haimana śasya* (autumnal crops), followed by *vasanitka muṣṭi* (spring harvest) in *Caitra* (March-April); and finally *vasantika śasya* (spring crops), followed by *vārṣika muṣṭī* (rains harvest) in *Jyeṣṭha* (May-June). Kauṭilya used *śasya* (crop) and *muṣṭī* (harvest), while similar meanings were conveyed by *vāpa* (sowing) and *pacyamāna* (ripening) by Pāṇini (fifth century BC) (Agrawala, 1963, pp. 194-205). A detailed description is given of the pre-planting treatment of seeds with various ingredients, like honey, ghee, cow-dung, pig's fat, etc. It is not clear as to what special properties are generated or in which way this process helps in sprouting of seeds. The idea may appear to be quite queer, but soaking with water or drew and drying in sun before putting in the seed bed may have some effect in producing seedlings. Transplantation of seedlings is hinted and use of cow-dung or cow-bones in the growth of plants is known for their manuring effect. Suitable plants have been listed for places with variable water contacts, like coastal strands, regions above high floods, lake beds, ridges, moist ground by the side of wells, etc. (2.24.22). The importance of water and moisture supply for the plant growth has, thus, been brought out by way of selecting plants suitable for growth in variable moisture-domains, selecting grounds and lands subjected to various process of submergence and moisture retentivity or hydrophilic character of soils. Crops to be sown during different times of rainy season have been indicated as: (1) in early rainy season — *śāli, vṛhī* (paddy), *kodrava, tila, daraka* and *varaka*; (ii) in mid-rainy season — *mudga, māsa, śimba*; (iii) in end of monsoon — *kusumbha, masūra, kulattha, yava, kalyaya, atasī* and *sarsapa*.

Crops and fruit trees suitable for lands receiving water from natural sources have also been indicated (Sen Sarma, 1998, pp. 120-2). Thus (a) *vallīphala*, that is pumpkin, gourd, water-melon, etc., (fruits produced by creepers) are suitable on land above the high water marks of rivers, sea-coast (*phenaghaṭaḥ*: where foam strikes the land); (b) *pippalī, mṛdivika, ikṣu* are suitable on land frequently inundated by flowing water (*parivahantaḥ*); (c) vegetables and root-crops on land at the fringe of water-reservoirs (*kūpaparyantaḥ* — reservoirs behind a dam, pond, etc.); (d) green vegetables on low-lying moist land (*haraṇiparyantaḥ*) and (e) fruits, like custard-apple, sweet-sop and other edible fruits, belonging to *Anonaceae* family, on high ground (*pali*). *Lavaṇa* should be taken to indicate the tropical genus of dicotyledons (*Amona Reticulata*, M.M. Williams, p. 898) which include these fruits.

And gradually from these identifications of crop-water relations, the discourse travels to the domain of irrigation.

Concept of irrigation

The concept of providing water to the crops when availability of natural water by way of precipitation and soil moisture is lacking, is immanent in the agricultural practices prescribed in accordance with the rainfall zoning, land characteristics, seasonability of crops, sowing time, etc., described under agriculture and agricultutral meteorology. According to the extent of availability of soil moisture at different time of the season, sowing of varieties of crops appropriate for that condition, and capable of extracting moisture for their survival and growth, has been indicated. This effective root zone, that is the depth of soil from which the plant draws water during its water sensitive stage of growth, varies from crop to crop. For example, there are shallow rooted crops, like rice, vegetables, potato, onion, etc., for which this depth is about 60 cm; for moderately deep-rooted plants like wheat, garden pea, carrots, etc., this depth is about 90 cm; depth is around 120 cm for deep-rooted plants like cotton, maize, citrus and fruit trees, pasture grass, etc., and about 180 cm for deep rooted plants like sugar-cane, tomato, etc. During the season when precipitation is less or absent, water levels in the soil gredually go deeper. Plants grown at that time require additional provision of water. Even during the precipitation season, water levels may go down below the root zones of plants due to deficiency of precipitation, or during spells of non-precipitation. Artificial means of supply of water then become necessary. This is in essence the underlying principle of irrigation. Irrigation requirement, thus, refers to the depth of water required to be provided for successful growth of plants, crops, etc., in a given time, exclusive of rainfall, ground water contribution and other natural resources. This is the modern understanding of irrigation, and that this aspect was well known then also is clear from the preceding discussions.

By irrigation practice is meant the induced means of providing water to the crops and plants at time when it is required and at place where it is required. It comprises the means, method and mechanism (the system of providing support and control of the use of resources by the establishment) to achieve this.

Importance of irrigation was highlighted at the outset of the treatise dealing with the establishment of settlements of the countryside and duties of other heads of departments. One of the primary duties of the official in charge of this act should be

> . . . . He should cause irrigation works to be built with natural water sources or with water to be brought in from elsewhere.
>
> — 2.1.20

The Director of Agriculture is expected to be conversant with 'water-divining' (2.24.1). The Superintendent of Pasture lands is supposed to attend to this task,

> In waterless regions, he should establish wells, water works and springs, also flower-and fruit enclosures. —2.34.8

While it would be the duty of officials incharge of agriculture or pasture land, they have been also advised to render all infrastructural supports to irrigation by way of roads, land, trees and implements (2.1.21). Attempts to create irrigational facilities should also be supported financially by exempting it from taxes, the exemptions varying according to the type of irrigation facilities created. Thus, newly constructed tanks and embankments would earn tax exemption for five years, the renovation work of earlier systems would be exempted from taxation for four years, clearing of weeds, etc. hindering flow, for three years and when dry land is brought under cultivation, tax exemption would be for two years (3.9.33). The king would, of course, levy tax on different forms of water abstraction, be it from public irrigation work or one's own arrangements, or from natural rivers, lakes, tanks or wells (2.24.18), and the king would be the sole owner of fish, ducks or green vegetables from the irrigation works (2.1.24). Irrigation water has not been allowed as a free-for-all commodity without price tag on it. Water cess for irrigational works is a necessary precondition for maintenance and creation of irrigational facilities, at the same time instilling the awareness of its value as a precious natural resource, to be used judiciously and not to be squandered away. In addition to levying tax on it, certain other restrictive and mandatory measures have been imposed, while accepting the right of individuals to the access of this resource. An individual can construct tanks, embankments, wells, other water works (3.9.33), lift water from rivers, lakes, tanks and wells (2.24.18) without obstructing customary water-courses in use or building thereon a non-customary structure (3.10.1). The individual's right to use has been endorsed by allowing the option to mortgage or of sale of such consttructed structures (3.9.34), or sale of water from his water works in exchange of share of the crops or horticultural products (3.9.35).

The practices of irrigation by individual should be such that they do not harm others, by indiscriminate or injudicious use of water. It would be the duty of everyone to take care and watch that while practising irrigation through a reservoir or channels, no damage is done to ploughing or seeds in another's field (3.9.27). The tanks constructed for the purpose should not also flood the adjacent fields by overflowing (3.9.29). Similarly, the flow of water from a tank constructed at higher level should not be prevented from flowing into a tank situated at a lower level, if the later is in use and has not ceased to function for three years (3.9.30). Such works made dry by emptying the tank would attract measures of punishment. Water works constructed but left unused for five years except in a distressing condition would be confiscated (3.9.31 and 3.9.32). The water works must always be kept in shape and use by repairing whenever needed, whether they are used by the owner, or leased out, given on rent, or by mutual agreement for use (3.9.36). Traditionally used charitable water-works or reservoirs meant for distribution of waters should not be tampered with, either in the form of sale or mortgage of the former, or drawing water from dams (reservoirs) out of turn (3.9.38 and 3.10.2). For the violation of all these stipulations, the offenders will have to pay fines, which would vary according to the nature and gravity of offence (2.1.22, 23, 3.9.27 to 3.9.30, 3.9.37, 38).

The aspects related to agriculture, viz., land utilisation, distribution, effects of climate on agricultural practice, science of irrigation, control of natural resources for their use, penalties for violation of injunction, etc., dealt with in *Arthaśāstra* bring out a well organised logical, political and bureaucratic organisational concept of higher order prevalent at that time. In the centre lies the concept of monarchical right of the properties, as source of revenue for the treasury. But that these properties — like land, water, agricultural and horticultural products are the things which demand careful management and preservation for deriving long-term benefits — are clearly revealed from the injunctions on rights, use, conservation, penalties for violations, etc. Several points emerge, which, it would be shown, bear so remarkable similarity with even the present-day concepts and practices. These may be discussed under the following headings:

(a) Perception of the agricultural science — agronomy, climate, land-use, irrigation, etc.

(b) Perception of the human efforts to overcome natural constraints, particularly in case of water resources.

(c) Objectives of human efforts on irrigation, as related to agriculture and economics.

(d) Recognition of the spheres of rights, freedom and liberty to the use of water resources; and of the necessity of introducing enforceable laws to ensure the appropriate use, avoidance of wasteful use; dealing with conflict and infringement on individual and community rights over common property as natural water, etc.

Though all these have been discussed in passing, some points are required to be elaborated in order to appreciate the modernism or thought process and actions programmes contained in *Arthaśāstra*.

The natural constraint in the form of paucity of rainfall and, hence, non-availability of water, has to be overcome by means of irrigation from two predominant sources — natural and artificial. Of these, the one with naturally flowing water is preferable to one which is created artificially (7.12.4).

Even amongst these two types of water works, the one which has more extensive area for sowing has to be preferred (7.12.5). The latter, obviously means that the one which has greater area on command, that is the area which is amenable to irrigation from that source, is to be preferred. From the descriptions of Megasthenese, it becomes clear that surface irrigation through canals used to be done exactly the way it is practised today. There used to be the main canal from the source of supply with a network and branches, leading throughout the command area, and the duty of the officials were to monitor the

> sluices by which water is let out from the main canals into their branches so that everyone may have an equal supply of it. . . .
> — McCrindle, *op cit.*, p. 86

A good deal of knowledge of hydraulics is necessary to maintain the system described. Perhaps these concepts clarify the understanding of purposes of irrigation — which are viewed from various angles. The agronomist's view of irrigation is to deliver water in accordance with crop needs. Other view emphasises the agricultural output made possible by providing irrigation, or in other words, suggesting that irrigation is a means of increasing the agricultural production. The government's view of irrigation is primarily a means of fostering overall economic development of a nation or a region within a nation. Social scientists may suggest that the purpose of irrigation is to make possible the overall well-being of people by ensuring adequate, secure and decent livelihoods. These disparate views arise out of the directness of the relationship of irrigational activities with agricultural production. One relationship defines the proximate purpose, that is the direct relationship of irrigational activities, as of supplying water to the crop's needs (agronomist's view). The other is indirectly related, in that the enhanced production by means of irrigation brings broader social purposes. Of course the achievement of the proximate or narrow purposes can lead to the realisation of broader purposes. This complexity can be resolved within a 'nested means/ends framework'. In this framework, a narrow purpose is seen as the means for achieving some specified end, which is the broader purpose within which the narrow purpose is nested. . . . This process continues until one arrives at the 'ultimate' purpose. (Mark Svendsen and Les Small, 1990). Irrigation is, thus, a means to an end, which may be

> defined as supporting a sustained level of agriculture that is more productive than would be possible under naturally rain-fed condition. — Svendsen, *op. cit.*

The policy on the use of water in irrigation, as emerges out of the various injunctions, can be stated as :

(a) Land and water of the country are monarchical properties, from which taxes were to be extracted, and exemption to be granted by the king on the basis of certain principles and traditional social practices.

(b) Water for irrigating the crop fields can be by preserving natural precipitation or by creating artificial storage.

(c) Water from natural courses may be used without interfering with natural flow or traditionally used water works.

(d) The right to use water is generally accepted, on condition of paying taxes for its abstraction, in whichever manner it is advantageous, the tax varying with the mode of abstraction.

(e) While the right to create water-works is endorsed, restrictions were imposed regarding their maintenance, of the structure and its functioning. Such water works are not supposed to function in such a manner as to cause any damage to the adjacent fields.

(f) Though the administrative machinery are to look after these aspects, including creation of irrigational facilities, the management of water resources appear to have been the function of individuals or community at large. Co-operative functioning, on respective share proportions had also been an accepted mode of operation.

The system of operation with water resources depicts also the people's right on their use, in recognition of their conventional and traditional rights, but with controls prescribed for violations or interference with the rights of others. It appears that the system conveys acute consciousness and awareness about the common good and community resource. The restrictions and legal obligations are intimately related to rights, which lose the meaning unless they are enforceable. To be enforceable, hence, right has to be a part of law.

It is interesting to note how these concepts compare with those prevalent in Indian constitutional and legal perceptions.

> The only right which exists in water in its natural state is the right of use. The right of property in water is usufructuary. Such a right is not in the ownership of corpus of water, but arises from the privilege to use it. The right of an individual to use water is entitled to the same protection as other forms of property. As the only ownership in waters is one of a right to use, this right can be exercised by an

> individual, a group of individuals or a legal entity. The rights exercised by the state over waters are not proprietary in nature but sovereign in character. From this flows the inherent right of the state to adminster or regulate the waters flowing within its territories. . . .
>
> — Alice Jacob and S.N. Singh, 1972, p.7

In India, irrigation practices are variously adopted, corresponding to the variation in climate, soil characteristics, topography, cropping practice, etc. These factors influenced the course of evolution of such practices from centuries. As a result, there are different irrigation acts for different states — which incorporate the various aspects, though in somewhat different manner, of crop water relationship, construction of irrigational works, their maintenance, role of state or individuals or groups of individuals in matters of use and supply of water, maintenance of water works, water cess, unauthorised irrigation, drainage and prevention of waterlogging, state's rights in natural waters, etc. There is hardly any aspect embodied in such acts which are not found in *Arthaśāstra*. Contemporaneity of these aspects and remarkable similarity of the concepts and perceptions contained in *Arthaśāstra* with those being practised in India, even elsewhere, compel to place it in the most advanced stage of understanding and knowledge in the fields of agriculture and irrigation.

The effect of the management of irrigation of the time is available from the accounts recorded by Megasthenese and Arrian. On the fertility of India,

> Megasthenese indicates the fertlity of India by the fact that soil producing two crops every year both of fruits and grains. Eratosthenes writes to the same effect, for he speaks of a winter and a summer sowing, which both have rains; for a year, he says, is never found to be without rain at both these seasons, whence ensues a great abundance, since the soil is always productive. Much fruit is produced by trees; and the roots of plants, particularly of tall reeds, are sweet both by nature and by coction, since the moisture by which they are nourished is heated by the rays of the sun, whether it has

> fallen from the clouds or been drawn from the rivers. . . .
>
> — McCrindle, *op. cit.*, pp. 54-5

Further

> India has many huge mountains which abound in fruit-trees of every kind, and many vast plains of great fertility — more or less beautiful but all alike intersected by a multitude of rivers. The greater part of the soil, moreover, is under irrigation, and consequently bears two crops in the course of the year. . . . The fact is, almost all the plains in the country have a moisture which is alike genial, whether it is derived from the rivers, or from the rains of the summer season, which are wont to fall every year at a stated period with surprising regularity. . . .
>
> — McCrindle, *op. cit.*, pp. 30-2

The most striking revelation of *Arthaśāstra*, from the treatment of water in the field of agriculture, as a sustaining means of plant and crop growth and survival and as a means of enhancing productivity, along with other influencing natural elements like soil, temperature, topography, etc., is the recognition of these elements as the 'resource', in the sense this has been defined. It is said that

> The word resource *does not refer to a thing or a substance but to a function which a thing or a substance may perform or to an operation in which it may take part*, namely, the function or operation of attaining a given end such as satisfying a want. In other words, the term 'resource' is our abstraction reflecting human appraisal and relating to a function or operation (*sic*).
>
> — Erich W. Zimmermann, 1951, pp. 6-7

Even today when the use of water resources is riddled with complexities, the core concept of its management veers round the above-quoted maxim. As late as 1993, the World Bank was evolving the policy of Water Resources Management on the basis:

> The proposed new approach of managing water resources

> builds on the lesson of experience. At its core is the adoption of a comprehensive policy framework and the treatment of water as an economic good, combined with decentralised management and delivery structures, greater reliance on pricing, and fuller participation by stakeholders. Decentralising the delivery of water services and adopting pricing that induces efficient use of water are key elements of sound water resource management. But, for a decentralised management to be effective, a supportive legal framework and adequate regulatory capacity are required, as well as a system of water charges to endow water entities with operational and financial autonomy and sustainable delivery of services.
>
> — World Bank Policy Paper, 1993

We have evidences of all these traditional knowledge, wisdom, foresight and policy framing acumen and mindset in Kauṭilya.

5

Environment

Natural and Civic

DIFFERENT aspects, which form the basic issues of today's environmental movements, often agitative, are also available implicitly or explicitly in a number of *sūtras* in *Arthaśāstra*. And surprisingly these are not in primitive form, but with fervent modernity. This can be realised when the intrinsic implications of the injunctions in *Arthaśāstra* are compared with the elements of environmental perceptions of the modern world, as revealed by the legal provisions, or the issues highlighted in environmental movement world over. Before proceeding with analysis in this line, it is necessary at the outset to clarify as to what is meant by environment, specially why this perception has been attributed as the pressing priority of human attention and action. This is no easy task, as is clear from the maze of definitions and approach to environmental problems by individuals, groups or schools. Some of the commonly used phrases in the various ways of defining environmental problems include such concepts as the indiscriminate human activities, too much of exploitation of the natural wealth associated with the industrial productions geared at profiteering motives; distortions and vulgarisation of the science and technology, socio-political differentials existing from one country to the other, from one continent to other continent and within the same society, so on and so forth. The fundamental perceptions of environment reflect and influence the emphasis and ways of tackling the problems.

> The perceptions, values, and preferences of numerous individuals and groups have to a very substantial degree

> determined how the natural world has been altered, what violence has purportedly been done to what existed, and what judgements can be passed on the ways in which human intervention and invention both have created the problems of contemporary society and offer the only possible solution for them.
>
> — Robert M. White, 1996, p. vi

The problems as

> presently defined — with respect to energy and food, but in most other instances as well-derive from the very specific scientific and engineering feats of recent decades.
>
> — White, *op. cit.*, p. vi

Yet, the older issues which occupied the minds of people of the other periods, still hold ground, particularly in the evolution of the human perception of environment. In this context, the definition propounded by Gordon Childe (1936) covers the entire gamut. It reads

> environment, of course, means the whole situation in which a creature has to live; it embraces not only climate (heat, cold, moisture, wind) and physiographical features like mountains, seas, rivers and marshes, but also factors such as food supplies, animal enemies, and in case of man, even social tradition, customs, and laws, economic status, and religious beliefs.

The definition apparently appears to be too simplistic and innocuous, but it covers the whole spectrum of human knowledge in classified disciplines, and at the same time it provides the orientation about the role of human being in his physical and social environment. These concepts are manifest in the injunctions inscribed in *Arthaśāstra*.

Environment related issues

The issues related to environmental concern may be classified as:

1. Natural set-up and resources: which comprise land, forest,

water, minerals, etc.

2. Biological environment, comprising animals, aquatic life.
3. Physical environment: comprising floods, droughts.
4. Human society: including hygiene, famines, civic responsibilities, civil rights, human right, etc.

Natural set-up and resources

Self-protection is the basic instinct of the organic world. Need for protection and preservation of the natural elements around man derives out of his perception of the potential of nature in the creative and productive process for his own development. This process changes and did change in the entire span of human history, with the change of attitudes, concept and invention of technology. In the primitive society, nature's gift was obtained in the form as it was offered. With time land, water, forest and other gifts of land — minerals, building materials, etc., received much attention for economic activities with these elements. The concept of materials, as resources, capable of yielding desired ouput with certain applied processes and activities, became operative and influencing force in shaping the social system.

In *Arthaśāstra*, land received much attention, its potentials identified and classified according to the concept of its being a resource. Patterns of utilisation, improvement, etc., of land were indicated, as described earlier. The various injunctions on land use, on do's and dont's, penalties and punishments specified for violation of these, rewards and incentives for proper use, etc., demonstrate the awareness developed at that time for the preservation of land as a resource. In assessing the excellence of a country, the *janapada*, the nature of land has been given much importance in *Arthaśāstra*. The ideal *janapada* is the one which was

> devoid of mud, stones, salty ground, uneven land . . . endowed with agricultural land with protected pastures . . . with farmers devoted to work. . . . —6.1.8

In addition to the natural quality of land, the value of lands depends on man's operation on it (7.11.9) that is how he makes use of it. It was perceived that

> cultivable land is better than mines because mines fill only treasury while agricultural production fills both the treasury and storehouses. —7.11.10-12

Man's labour and activities being the main forces of converting the inert or neutral material and substance into productive resource, it was realised that

> As between land with the support of a fort and one with the support of men, the one with the support of men is preferable. — 7.11.23

because

> without men, like a barren cow, what could it yield? —7.11.25

Human endeavours have been appropriately emphasised:

> one trusting in fate being devoid of human endeavour, perishes, because he does not start undertakings or his undertakings have miscarried. —7.11.34

But the danger inherent in aimless or anarchic activities was also foreseen,

> one doing whatever pleases him does not achieve anything — 7.11.3

and . .

> he is the worst of them all —7.11.36

Kauṭilya's concerns for natural resources are cloaked in different *sūtras* dealt by him under apparently unrelated topics. It is only when the innerlying spirit and nuances of these are conjointly studied, that his acute awareness for the nature's preservation becomes clear. While explaining the traits which contribute to the 'excellence' of 'constituent elements' of the State (that is the king, the minister, the country, the fortified city, the treasury, the army and the ally), the characteristics which determine the excellence of the country have been described (6.1.8). And when conditions are otherwise or opposite to such considered as excellent for any of the elements, it has been termed and considered as calamity (8.1.3). Unlike the common meaning of the term,

> Kauṭilya uses the word 'calamity' (*vyasana*) in the precise sense of any event which weakens any constituent element of a state, hereby preventing it from being used to its full potential . . . . — Rangarajan, p. 117

A calamity of a constituent, of a divine or human origin, may be derived from either ill-luck or wrong policy. Undoubtedly from these perceptions emanated the injunctions to be imposed on acitivities otherwise harmful to land use, agricultural practices, preservation and development of forests, for their porducts and place for sheltering animals. The integrated approach for preservation, protection and development of land, forest, animals and vegetation is to be found in a number of *sūtras*. Judicious use, observance of non-destructive and non-derogatory practices, by adhering to appropriate technology and avoiding such human actions as might bring about deterioration, have been advocated for preservation of environment.

Appropriate use of resources and administrative control to restrain over exploitation, constitute the prime paradigm of environmental protection, and all these are to be found in the way land use, agricultural practices, etc., were mandated. These have been discussed earlier in the relevant section.

Animal welfare and protection of bio-diversity

Edicts and proclamations in *Arthaśāstra* bear ample testimony of Kauṭilya's perception and concern about the living creatures — animals, both wild and domestic, and plants and vegetation.

Duties common to all are: abstaining from injury (to living

> creatures), truthfulness, uprightness, freedom from malice, compassion and forebearance. —1.3.13

While general proclamations were directed to the common people, high-ranking officials used to be appointed with specific responsibilites and authority to execute them. Penalties and punishments were specified for violations of edicts. Thus, amongst the high officials, special positions were occupied by Director of Forests, Supervisor of Animal Slaughter, Superintendents of Cattle, Horses, Elephants and Pastures. These officials were assigned with duties of protection of wild life, ensuring proper rations for animals, regulating grazing, preventing cruelty to animals, preventing poaching of wild animals, ensuring proper care of domestic animals, etc., and imparting penalties and punishments specified for violation of the edicts.

Animals, particularly horses and elephants were given formidable importance, for their use in warfare. Cattle (cows, bulls and buffaloes), goats, camel, donkeys and pigs, in the crown herd, were to be meticulously looked after, for which a Chief Superintendent of Crown Herd used to be appointed. Even individuals whether owning domestic animals or not, were supposed to follow norms about their dealing with the animals. These atitudes may appear at the outset to have been evolved from utilitarian considerations, but edicts would dispel that they were totally derived of mercenary outlook. The official incharge of disposing non-agricultural land was supposed to establish an animal park in conformity with the grain of the land, where all animals are welcomed as guests and given full protection (2.2.4). Similar forests for elephants should be established providing due protection to the animals (2.2.6-7). Though such sanctuaries are meant to rear the elephants which could be captured for use in the war or other uses, indiscriminate caputre or killing were prohibited. Any one caught of killing an elephant was to be executed (2.2..9). Killing or injuring animals and bird which were declared as protected species, in reserved parks and sanctuaries, were proclaimed as punishable acts (2.26.1). Lists of protected animals, fish, birds were drawn up. For killing of animals, birds, etc., whose slaughter or capture is not customary, heavy punishment was to be imposed (2.26.2). Indiscriminate killing, even customarily sanctioned, of such species would not be allowed (2.26.3). Among animals customarily slaughtered for meat, killing of calf, bull and milk cow was prohibited (2.26.10-11). Animals within the sanctuaries and parks, found to be harmful or dangerous, were to be taken out of the sanctuaries and then killed (2.26.14).

Strict instructions were given for treating the domestic animals belonging to individuals, village communities or to the king. Village headman was made responsible for preventing cruelty to animals. Stray protected animals or those from reserve forests found grazing at places where they were not supposed to be, should be driven off without hurting them (3.10.31). Stray cattle should be driven off with a rope or whip. People trying to attack such cattle should be prevented from doing so by all available means and those found to hurt the cattle were to be fined (3.10.33-4).

The herdsmen would be responsible for caring for the young, the old and diseased cattle and treat them when required. They should graze them in safe forests, after the thieves, wild animals and other kinds of predators have been removed. The pastures suitable in different seasons should be identified for grazing. Timid cattle should be provided with tingling bells to frighten snakes and wild animals; they should be taken to safe watering locations through easily moveable roads and which should not be infested with crocodiles (2.29.20-3). Diseased cattle, elephants and horses should be treated by physicians, proper food to be given to them, for which the quality and quantity of ration required to be given were specified. Details of care, training, treatment of horses and elephants, milking of cow, etc., have been indicated (Book 2, chapters 29, 30 and 31).

The following depicts the administrative injunctions related to the animal welfare and their protection. These are the acts, which were considered as punishable, ranging from pecuniary fines to death, mutilation of limbs, nose, etc.

Acts, which were considered punishable, were (Rangarajan, pp. 322-3):

(a) Trapping, injuring or killing of protected species, animals in sanctuaries (2.26.1).

(b) Above offence committed by householders (for their personal use) (2.26.1) or game keepers and sanctuary guards allowing these offences happen (2.26.6-2.29.16).

(c) Trapping, injuring and killing of fish and birds whose slaughter is not customary (2.26.2) or deer and animals whose slaughter is not customary (2.26.2).

(d) Killing or torturing to death a calf, bull or milk-cow (2.26.11) or castrating the male of a small animal used for breeding (3.20.17).

(e) Injuring animals with sticks or causing bleeding wounds to animals (3.19.26-7).

(f) For allowing the bulls of a herd to fight each other or letting horned or tasked animals fight and kill one another (2.29.38; 4.13.19).

(g) Riding or driving away temple animals, a stud bull or a cow not yet calved (4.13.20).

(h) Killing of small animals like cocks, cats, dogs, pigs, etc., (4.10.2).

(i) Killing of small animals useful for its milk, hair, for riding or for stud (4.13.21).

Much emphasis was given on the rearing, care and protection of forests, gardens, orchards — these being considered as sources of revenue, besides being of recreational spots like parks. The countryside suitable for establishing animal parks or forests, as suitable resorts of elephants were classified.

The king is supposed to protect old-produce forests, elephant forests, irrigation works and mines that were made in ancient times and also should start new ones (2.1.39). The irrigation works, by Kauṭilya's definition, encompass flower gardens, fruit orchards, vegetable gardens, wet-crop fields and sowing of roots (2.6.5). Forests should be established one each for the products of forests, due care is to be taken for their up-keep, and factories for goods made from these products would have to be started (2.2.5).

Cutting of shoots of flowering or fruit bearing trees, or trees which provide shade were prohibited. The trees were not to be damaged either by cutting of small branches, or stout branches, or trunks or they must not be uprooted. For violation of each of these conditions, that is for harming the trees by cutting any part of a tree, pecuniary fines of variable amounts, depending on the types of injury to the trees, as described, were to be imposed on the offender (3.19.28). Any one doing damage even to bushes and creepers bearing flowers or fruit or providing shades were liable to be penalised. Provisions of penalties were made for damage done to trees in holy places, penance-groves and cremation grounds (3.19.29). Heavy fines were also to be levied for causing harm to trees at the boundaries, sanctuaries, and also to the prominent trees.

Civic responsibilities for ptotection of vicinal environment

In so far as 'environment' is considered as the 'sum total of all conditions and influences that affect the development and life . . . ', the human life, his behaviour, concepts, attitudes towards fellow members, community, outward world, etc., in short all the highest forms of cultural patterns, are to be considered within the broad spectrum of this definition. Human character being as it is, his intensity of concern generally peters outward from self to family to neighbours and lastly towards the greater world, outward beyond his immediate intimate spheres. His capacity and self-education to develop concern, comity, co-operation and attitude to live and let live amongst his fellow members and community determine his level of concern and care of the environment around him. Civic duties and rsponsibilities of every human being, perhaps, are the basic ingredients which offer the initial lessons and generate the higher philosophy of life and living within the wider world comprising all animate and inanimate environment.

When injunctions and directives in respect of these aspects are analysed, it becomes clear that these deal not only with the mundane affairs, like personal and community health and hygiene, cleanliness of the surroundings, avoidance of pollution of water, etc., but expound also the inner lying philosophy and attitude expedient to be followed and practised as a good neighbour and protector of nature.

Harmonious, peaceful and dignified life of people was considered as most desirable and this mode of life was to be valued more than wealth and other worldly acquisitions.

> Making enemies is a greater evil than loss of wealth.
>
> — 8.3.18

In dealing with people harms that could follow from anger, lust and greed are to be properly understood and hence controlled.

> Anger makes one the object of hatred, creates enemies and brings suffering on oneself. Excessive greed and lust bring about humiliation, loss of wealth and association with undesirable persons . . . [while both have had consequence] being hated is worse than humiliation; whereas one humiliated is held in thrall by his own people and by enemies, a hated one is destroyed.
>
> — Rangarajan; p. 137, tr. of *sūtras* 8.3.13 to 8.3.22

The edicts meant for individual and community behaviour are in conformity with above perception.

It is desired that no one should behave in a manner likely to cause any harm to the immediate neighbourhood (3.20.15). To avoid any inconvenience of confrontation, the houses should be constructed in conformity with the demarcated boundaries to be laid with sufficient space between two houses (3.8.3-5); parking place for carts and quadrupeds, fire place, water storage arrangements, grinding mill or corn thrashing devices, etc., should be placed sufficiently away from a neighbouring house (3.8.11); any obstruction by way of ditch, outside projecting stair case, ladder, water channel, etc., are to be avoided (3.8.21); and for effecting such arrangements house-owners, by mutual agreement, may get things done as desired, and should avoid what is undesirable (3.8.18).

The people are supposed to perform duties towards community

living.

> Everyone shall contribute his share to the building of common facilities. No one shall obstruct or prevent the lawful use of such facilities by others in the neighbourhood. Such facilities shall not be destroyed.
>
> — 3.8.26-7

> Sheds, courtyards, latrines, fire places, places for pounding grain and all open spaces are to be used as common property. — 3.8.28

> No one shall damage plants and trees in city parks, sanctuaries and holy places and cremation grounds, particularly those which bear fruits or flowers or provide shade. — 3.19.28-30

Anarchic lifestyle, indiscriminate modes of waste disposal, ignorance, negligence or disregard of the consequences of careless activities which result in pollution of air and water, and bring about health hazards, were prudently recognised. Means to avoid all these eventualities were indicated in edicts, for observance by every citizen, for violation of which, penalties and punishments were proclaimed and for implementation of the dictates, officials were engaged. Thus, no one should throw dirt on the road, or allow mud and water to collect there, which would block the traffic movement; no one shall pass urine or relieve himself in and around holy places, water bodies, a temple, a royal property, unless forced by the effect of medicine, illness or sudden fear; no one shall dispose off dead bodies of domestic pets, like a cat, a dog, carcass or donkey, camel, mule, horse or cattle or a serpent killed, anywhere inside the city (2.36.26-30). Human dead bodies shall be cremated or buried only in designated grounds. Corpses are not to be thrown anywhere outside, and should be carried through prescribed routes; violation of these are punishable, both for those who do not follow the dictates or those (like gate-keepers) who allow these (2.30.31-3). Water course, water wells and dung-hill should be placed only at specified places suitable for households (3.8.6), free flow of water in channels during rains must not be obstructed (3.8.23). These precautions are obviously meant to prevent air and water pollution and ensure cleanliness of the surroundings.

It is interesting to observe some of human behaviour considered as a subject common to civil and criminal laws in India. It is dealt under the broad caption of 'nuisance', the perception of which is, perhaps, one of the concepts of civilised life and which entered into the law of torts from ancient times (injunctions in *Arthaśāstra* are comparable).

> Even today this law of torts is available as a remedial assistance to the sufferer, by way of monetary compensation or granting order to the wrongdoer requiring him to do or to refrain from doing some act (injunction).

Nuisance can be divided into public and private nuisance.

> An act or omission which materially affects the reasonable comfort and convenience of section of people is a public nuisance. The sorts of activity that amount to environmentally damaging public nuisances include . . . carrying on offensive traders, rubbish dump, cess pit or other collection of filth that affect the health and habitability of a locality.
>
> Private nuisance affects only an individual or particular individuals, that is, those who have particularly become subject to the nuisance complained of.

Several nuisance acts were in vogue in metropolitan cities, cantonments and towns based on the standards of hygiene and sanitation.

Indian Penal Code provides punishment clauses for such acts as (i) negligent acts by individual(s), which are likely to spread infection of diseases dangerous to life, (ii) adulteration of food or drink intended for sale[1], (iii) adulteration of drugs, (iv) act of polluting water, (v) offence of making atmosphere noxious to health.

Public nuisance is considered a crime, though it can also be a tort in certain circumstances, while the private nuisance is always tortious (Chaturvedi *et al.*).

Calamities

Natural and man-made calamities, which can afflict population have been listed. Natural calamities are considered as acts of God and prevention of these has been prescribed through worships, oblations and recitals of benediction. Side by side, the duties of citizens, kings and his officials prescribed at the time of calamities indicate realistic and scientific approach to prevent, abate or adjust the calamitous situation to safeguard and save the population. Thus, during floods, which normally happen during the rainy season, the population of the places situated by rivers and water bodies should shift to areas above flood levels. There should be elaborate preparedness, by way of collecting wooden planks, bamboo, boats, skin-bags (inflatable), canoes, tree-stems, rope braids, hollow gourds — so that these could be utilised for safe transportation away from flooded areas and for rescue operations. Strict instructions are there for attending to rescue operation, and anyone found to be inoperative without valid reason was to be punished (4.3.6-9).

Drought has been identified as a greater calamity than excessive rain:

> The loss of the crops is a greater evil than the loss of sowings, as it involves loss of efforts, draught (a greater evil) than excessive rain, as it involves loss of livelihood. —8.2.25

---

1. Comparable Sutras in *Arthaśāstra* are:

   2.26.7: (Traders) shall sell meat without bones, of deer and beasts freshly killed.

   2.26.8: For (meat) containing bones, they should give a compensation for loss.

   2.26.12-13: They shall not sell (meat that is) swollen, without head, feet and bones, foul smelling and (of a) naturally dead (animal). Otherwise, there shall be a fine of twelve panas.

From the propitiation prescribed, it transpires that all the gods to be worshipped for alleviating drought, like Indra, the Ganges, the lord of the sea, are related to water. Though no appropriate measure to fight drought has been prescribed, yet the above indicates that measures for provision of water are called for during drought.

Fire, famine, disease in epidemic form are the other calamities identified as greatly affecting people. Floods, it is perceived as more pervasive than fire, since the former carry away hundred of villages, while the latter burns at most a village or its part (8.4.4). Similarly, disease is considered to afflict only one region and remedies can be found for it, while faminen afflicts the whole country and affects livelihood of vast population (8.4.7). Extensive relief measures are to be provided during the famine — seeds and food staff are to be stored and afflicted people should be favoured with food, or provided food — for work, by way of construction of forts, water-works, or should be sheltered with allies or in regions where crops have grown, or settle along the sea, lake or tanks and resort to fishing, hunting or cultivation along the water works (4.3.17-20).

Prevention from fire

Prevention, preparedness and expeditious measures prescribed during natural calamities demonstrate the mental faculty and prudence of understanding the causality of such calamities and accordingly the measures called for. Similar understanding is apparent from the edicts related to the cause of fire and measures to be taken.

> In the summer, citizens shall take appropriate precautions against fire. They shall not light fires during the (two) middle quarters of the day; if food has to be cooked during this period, it shall be done outside the house.
>
> Every households shall provide five pots, a big jar, a trough, a ladder, an axe (for chopping down pillars and beams), a winnowing basket (for fanning away the fire), a hook (to pull away burning parts), a hooked rake (for pulling away the thatch) and a skin bag.

> During the night, householders shall stay near the front doors of their houses and shall not collect together.
>
> If a house catches fire, every occupant, owner or tenant, shall take immediate steps (to put it out).
>
> — Tr. of *sūtras* 2.36.15-18, 21, 23 by Rangarajan, *op. cit.*, p. 370

The City Superintendent was under strict instruction to watch and implement the instructions.

> Constant inspection of places, supplying water, roads, water courses, covered paths, ramparts, parapets (and other) fortifications, and the guarding of what is lost, forgotten or has run away (are the duties of the Superintendent). — 2.36.42

The king has also to follow precepts in this respect, that is in matters related to environment.

> It is in the interest of the prosperity of the country, a king should be diligent in foreseeing the possibility of calamities, try to avert them before they arise, overcome those which happen. . . .
>
> — Tr. of *sūtras* 8.4.50 and 8.5.21 by Rangarajan, p. 116

Contemporaneity of Kauṭilyan concept of environment

More than any other subjects, discussed earlier, Kauṭilya's precepts, edicts and proclamations on environmental issues not only compare with the most modern concepts, but reveal a highly sensitive and transcendental mind. History proves that human concern for the inanimate world, need for its protection and preservation assumed a concrete shape in thought process and action programme only recently, after gradual sensitisation of human mind to these ends following the Industrial Revolution. Historians charted the trend of the social systems and their evolution in relation to the productive process in which nature was once treated 'as little more than a resource to be tapped and as a sink into which to dump their wastes'. This unidirectional concept had its evil effect, even in contributing to the destruction of a civilisation.

The 'emergence, in stages, of the capitalisst world system, which began in Europe in 1400s' ultimately culminated, through complex political and economic changes, to

> the next great stage of capitalism's advance, the Industrial Revolution, which took place in Europe in the late eighteenth and early nineteenth centuries and led to a rapid increase in the scale and intensity of production and to the development of a set of divisions that are at the core of our understanding of the modern world: the division between economy and nature, between capital and labour, and between centre and periphery.
>
> —John Bellamy Foster, 1995

On this basis it is rightly recognised that the cause of the degradation of nature lies within the system of a society. The earlier concept of nature continued to take the front-stage in capitalist production, plundering of natural resources became the mainstay in the production process catering to a world craving for more, greater and higher consumption. Backlash of this became starkly apparent from wider and large-scale pollution and degradation of the biosphere, the ill effects affecting the mankind and portending long-term irrevocable disastrous end results.

This situation generated the consciousness about the worldwide problem of maintaining a safe environment for human existence, and only recently, the United Nations Organisation held in 1972 its conference on Human Environment (Man and Biosphere) in Stockholm. With the adoption of its resolutions, successive global conferences followed on various issues related to environment, and the concerted move on environment-related activities began. Some of the principles proclaimed during the conference are so relevant and similar to the edicts of *Arthaśāstra* that it justifies to reproduce them.

PRINCIPLE 2

The natural resources of the earth, including the air, water, land, flora and fauna and especially representative samples of natural ecosystems, must be safeguarded for the benefit of present and future generations through careful planning or mangement, as appropriate.

PRINCIPLE 3

The capacity of the earth to produce vital renewable resources must be maintained and, wherever practicable, restored or improved.

PRINCIPLE 4

Man has the special responsibility to safeguard and wisely manage the heritage of wildlife and its habitat, which are now gravely imperilled by a combination of adverse factors. Nature conservation, including wildlife, must therefore receive importance in planning for economic development.

PRINCIPLE 8

Economic and social development is essential for ensuring a favourable living and working environment for man and for creating conditions on earth that are necessary for the improvement of the quality of life.

PRINCIPLE 9

Environmental deficiencies generated by the conditions of underdevelopment and natural diasters pose grave problems and can best be remedied by accelerated development through the transfer of substantial quantities of financial and technological assistance as a supplement to the domestic effort of the developing countries and such timely assistance as may be required.

It is interesting to note the methods prescribed in *Arthaśāstra* of counteracting famine, a disastrous natural calamity, in *sūtra*s 4.3.17 to 4.3.20 — which may be summarised as: (i) distribute to the public, on concessional terms, seeds and food from royal stores; (ii) undertake food-for-work programmes such as building forts or irrigation works; (iii) commandeer for public distribution of private stocks of food; (iv) seek help from friendly kings; (v) move the entire population to region or country with abundant harvest or near the sea, lakes or rivers; (vi) supplement the harvest with additional cultivation of grain, vegetables, roots and fruits, by fishing and hunting deer, cattle, birds and wild animals (tr. by Rangarajan, *op. cit.*, p. 130).

PRINCIPLE 13

> In order to achieve a more rational management of resources and thus to improve the environment, states should adopt an integrated and co-ordinated approach to their development planning so as to ensure that development is compatible with the need to protect and improve environment for the benefit of their population.

PRINCIPLE 15

> Planning must be applied to human settlements and urbanisation with a view to avoiding adverse effects on the environment and obtaining maximum social, economic and environmental benefits for all. . . .

PRINCIPLE 17

> Appropriate national institutions must be entrusted with the task of planning, managing or controlling the environmental resources of states with a view to enhancing environmental quality.

Following the Stockholm Conference and adoption of principles, almost all the countries of the world framed environmental protection and management Acts and Rules in conformity with the national and international requirements. A cursory glance into such Acts and Rules framed in India and provisions made within their framework would provide almost a global picture (Chaturvedi, 1996).

The Constitution of India initially did not have any direct provision for environmental protection. Following the Stockholm Conference and ever growing awareness of the environmental crisis within and outside India, the Constitution was amended in order to insert direct provisions for protection of environment (42nd Amendment). Direct provision, accordingly, was inserted first in the part dealing with the Directive Principles, as Article 48A, which reads,

> The State shall endeavour to protect and improve the environment and to safeguard the forests and wildlife of the country.

and second in the Fundamental Duties of the Citizens, earmarked in Article 51A, placed in part IVA of the Constitution, in which clause (*g*) provides:

> It shall be the duty of every citizen of India to protect and improve the natural environment including forests, lakes, rivers and wildlife and to have compassion for living creatures.

It is tempting to quote some of the provisions of various Indian environmental Acts, if only for their amazing similarity of forms and contents with *Arthaśāstra*.

A FEW PROVISIONS IN THE WATER (PREVENTION AND CONTROL OF POLLUTION) ACT, 1974

(a) No person shall knowingly cause or permit any poisonous, noxious or polluting matters determined in accordance with such standards as may be laid down by the State Board to enter (directly or indirectly) into any (stream or well or sewer or land).

(b) No person shall knowingly cause or permit to enter into

> any stream any other matter which may tend, either directly or in combination with similar matters, to impede the proper flow of the water of the stream in a manner leading or likely to lead to a substantial aggravation of pollution due to other causes or of its consequences (comparable *sūtra*s 2.36.26-30, 2.36.43).

THE FOREST (CONSERVATION) ACT, 1980

The object of the Act is to 'provide for the conservation of forests' and section 2 is the vehicle for the object. The Act has been passed with a view to checking deforestation. Deforestation causes ecological imbalance and leads to environmental deterioration. With this object in view, the section has made it obligatory upon the State Governments to obtain prior permission of the Central Government for (a) dereservation of forest, and (b) use of forest land for non-forest purposes (comparable *sūtra*s 2.2.3-7, 2.1.39).

THE WILD LIFE (PROTECTION) ACT, 1972

The Act directs that

> No person shall destroy, exploit or remove any wildlife from a sanctuary or destroy or damage the habitat of any wild animal or deprive any wild animal of its habitat within such sanctuary except under and in accordance with permit granted by the Chief Wild Life Warden and no such permit shall be granted unless the State Government being satisfied that such destruction, exploitation or removal of wildlife from the sanctuary is necessary for the improvement and better management of wildlife therein, authorises the issue of such permit.

Such permits are issued in cases where (i) any wild animal has become dangerous to human life or property (including standing crops), (ii) where an animal is so disabled or diseased as to be beyond recovery.

The Act provides for appointment by the Central Government of different officials like Director of Wild Life Preservation, Assistant Directors and such other officers and employees as may be necessary. Similarly the State Governments can appoint Chief Wild Life Warden, Wild Life Wardens, etc. (comparable *sūtras* 2.26.1-6, 2.26.14, 2.17.1, 2.31.1).

The Act also prohibits picking, uprooting, damaging, destroying, acquiring or collecting specified plants from any forest land or any area specified by notification, or possessing, selling, offering to sale, or transferring by way of gift or otherwise, or transport any plant, whether alive or dead or part or derivative thereof, except of grant for special purposes (comparable *sūtras* 2.1.39, 2.2.5, 2.6.5, 3.19.28-30).

That age of transcendence

The United Nations General Assembly, as late as 1982, adopted 'The world charter for Nature', declaring that

> (a) Mankind is a part of nature and life depends on the uninterrupted functioning of natural systems which ensure the supply of energy and nutrients; (b) Civilisation is rooted in nature, which has shaped human culture and influenced all artistic and scientific achievements, and living in harmony with nature gives man the best opportunities for the development of his creativity, and for rest and recreation.

This worldwide consciousness and understanding of the oneness of man and nature by all the nations, irrespective of differential social systems, industrial status, technological achievements and productive outlook, have been brought about by force majeure through trials, tribulations and traumatic experiences. Voices, feeble though, were raised earlier also against human action causing degradation of the nature, but it is only during the post-Second World War period, that the conservationist ideas started crystallising into action programmers. It had a modest beginning in the United States in mid-nineteenth century following the publication of 'Man and Nature' by George Perkins Marsh in 1864, almost simultaneously with the publication by Karl Marx's *Capital*, a criticism of the industrial capital and related issues. The major concern leading to the growth of conservation movement, was the rate of extinction (of wildlife, forests, grassland, etc.) (Foster, *op. cit.*, pp. 72-4). Extinction of biotic life was directly connected to the market principles where natural resources were indiscriminately plundered for profiteering. But then, this destructive attitude is deeply-rooted in the man's philosophy and perception about himself and nature.

For centuries, perhaps up to the time of Copernicus in the sixteenth century, anthropocentric philosophy of the universe dominated the Western philosophy, and the basic nuance of it continued even thereafter till our own time. That is if

> even in the present we usually take for granted that 'here we are as human beings', and that some of us refer ourselves, scientifically, to the rest of the world around us. The Universe then appears to be our environment, the human habitat. The approach is consistent with the anthropocentric dualism of being and having, namely to be human and to have everything else at our disposal.
>
> — Klaus Michael Meyer-Abich, 1996, pp. 212-13

And this attitude gained ground with the ushering of the capitalist system, and was supported by men of science and culture.

> Nothing so reflected the imperial spirit of the age as the outlook of the seventeenth century science, which saw humanity as engaged in a war for the domination of nature.

Francis Bacon, belonging to that period, observed that conquest of nature constitutes 'the real business and fortunes of the human race'. He further observed that 'By art and hand of man', the nature should be 'forced out of her natural state and squeezed and moulded'. (John Bellamy Foster, *op. cit.*, p. 41).

That was the perception about nature, and which influenced the human action on and with the elements of nature to convert them as resources for use by mankind, in the sense Zimmermann defined resource (discussed earlier).

Recognition of material elements of nature as useful for human society and livelihood was not new or particular of the time beginning with the advent of profit-motivated system of production. The ascent of civilisation is measured in reference to the use of natural resources (in the common parlance of the word, meaning the materials). But what has been noticed, as discussed earlier, is the change in the approach to such use. Technological development is correlatable with such approach — Technology influences and is influenced by the social praxis and attitudes.

The social system, as revealed from *Arthaśāstra*, is that of a Tributary Society. The prime resource — land — and its 'produce was exercised through the extraction of tribute from peasant producers', which has been considered as a 'Tributary Society' (Foster, p. 36). The wealth of the king depended on the flow of the tributes. But extraction of such tributes was never stretched to the point of exploitation — either of the land's productivity or of the producers. Care for the land and associated care for the producers can be observed to be the hallmark of the period. This aspect has been clearly reflected in the contemporary accounts available in the written form. The accounts when interpreted in the context of human perception and attitude, it transpires that the attitude towards nature acquired a moral meaning. Or in other words,

> the independent value of living nature is being recognised in practice and that it is being incorporated into the field of morally responsible behaviour.
>
> — Bakshtanovsky *et al.*, 1989, p. 218

One of the accounts relates,

> . . . . there are usage observed by the Indians which contribute to prevent the occurrence of famine among them; for whereas among other nations it is usual, in the contests of war, to ravage the soil, and thus to reduce it to an uncultivated waste, among the Indians, on the contrary, by whom husband men are regarded as a class that is sacred and invisible, the tillers of the soil, even when battle is raging in their neighbourhood, are undisturbed by any sense of

> danger, for the combatants on either side is waging the conflict make carnage of each other, but allow those engaged in the husbandry, to remain quite unmolested. Besides, they neither ravage an enemy's land with fire, nor cut down its trees.
>
> — McCrindle, *op. cit.*, p. 32

While discussing about the natural environment in terms of the relation of nature and man, the concept of the latter about the former, emergence of heliocentric philosophy of Copernicus in the sixteenth century AD, in which man is seen as an integrated but infinitesimal part within the 'great garden of universe', and the present-day pleonastic encomium for the emerging in holistic universal consciousness and concern, it hardly escapes notice that similar attitude was entrenched in the Indian society as depicted in *Arthaśāstra* in fourth century BC. It existed within a higher form of culture, environment was perceived not as it appears but more in the sense which has euphemistically been called as

> co-natural, instead of the 'environment' with its unfortunate anthropocentric connotation.
>
> — Klaus Michael Meyer-Abich, *op. cit.*, p. 214

The principles, philosophy and imperative injunctions in *Arthaśāstra* were meant to have the motivating power, guiding and regulating peoples' behaviour and their co-natural stand in life.

And to seek the genesis of this form of transcendental approach to life and nature, one has to look through the social, political and religious history of that time, that is when Kauṭilya succeeded in destroying the Nanda dynasty and installed Candragupta Maurya on the throne of Magadha, immediately after Alexander's invasion of India. The period was partaking in the changes in various spheres, particularly in the pattern of thought process which began in India in the middle of the first millennium before Christ (BC). The Indus civilisation having declined by the second millennium BC, the new culture grew in plains of the Ganges towards the eastern extremity of India.

Magadha and the countries east of it lay almost outside the

> pole of Vedic culture. Perhaps the Vedic polity, along with Vedic culture, had spent its force, and it was now the turn of the sturdy peoples of the east, comparatively unaffected by the Aryan invasion, to play their part. When the Puranas lament the wholesale destruction of the kshatriya power by the sudra king Mahapadma Nanda, they seem instinctively to refer to the end of one epoch and the rise of another. The laxity of social restrictions imposed by the orthodox Brahmanical culture and the universal aspect of Buddhism and Jainism which found a congenial home in Magadha must have considerably widened the political outlook of this region and contributed to make it the nucleus of a mighty empire. — Mukherjee, R., 1953, pp. 35-6

The first millennium BC was designated as the 'axial time' and

> it was then that our intellectual, moral, and religious civilisation was born and that the foundations were laid on which we continue to build, despite differences in the superstructures we have erected and go on erecting.

Greek and Hebrew civilisation took shape at that time. And

> Surprisingly enough, other civilisations, practically without contact with, and certainly not influenced by, our nascent thought systems show astonishingly parallel developments.
> — Weil, *op. cit.*, p. 21

The most formidable influence on the thought process and social system came from Buddhism and Jainism in India. For a number of reasons, which comprised simplicity of doctrinal framework, expressed in intelligible common people's language, freedom from stringent rituals, honour and dignity professed for each man, irrespective of cast or profession, pursuance of simple but practical methods in order to arrive at truth, and so on, Buddhism was widely embraced, even becoming the royal religion. No doubt that Buddhist doctrines and philosophy strongly influenced the thinkers of those days. *Arthaśāstra* is replete with such comparative precepts. Buddhism propounds that to remove ignorance, greed, attachment, etc., one is required to grasp four noble truths (*ārya satyas*), viz., *duḥkha* (worldly miseries), *samudaya* (the thirst, attachment, etc., being the cause of worldly existence), *nirodha* (ending of worldly existence by destruction of thirst, attachments, etc.), and *mārga* (the path or ways of destruction of thirst, etc.). The *mārga* has eight-fold ways, viz., right speech, right action, right means of livelihood, right exertion, right-mindedness, right meditation, right resolution and right point of view. All these lead to physical control (*śīla*), mental control (*citta*) and intellectual development (*prajñā*) (Datta, Nalinaksha, 1953, p. 370).

Kauṭilya considers six traits of human character as enemies against self-control, discipline and knowledge. On the subject of renouncing these enemies, *Arthaśāstra* states —

> The sole aim of all branches of knowledge is to inculcate restraint over the senses. Self-control, which is the basis of knowledge and discipline, is acquired by giving up lust, anger, greed, conceit, arrogance and foolhardiness. Living in accordance with the *śāstras* means avoiding over-indulgence in all pleasures of (the senses, i.e.) hearing, touch, sight, taste and smell. A king who has no self control and gives himself up to excessive indulgence in pleasures will soon perish, even if he is the ruler of all four corners of the earth.
>
> — *Sūtras* 1.6.1-4, tr. by Rangarajan, *op. cit.*, p. 144

In contrast, a king who has these qualities can be called *a rājarṣhi* (a king wise like a sage):

- has self-control, having conquered the (inimical temptations) of the senses.
- cultivates the intellect in association with elders . . .
- is ever active in promoting security and welfare of the people.
- ensures the observance (by the people) of their *dharma* by authority and example.
- improves his own discipline by (continuing) learning in all

branches of knowledge.

- endears himself to his people by enriching them and doing good to them.

Such a disciplined king should:

- keep away from another's wife.
- not covet another's property.
- practise *ahiṁsā* (non-violence towards all living things).

— Tr. of *sūtras* 1.7.1-5 by Rangarajan, *op. cit.*, p. 145

Relevant to the issue under discussion, that is the growth of awareness for protecting and preserving the natural and societal environment, is the emphasis on non-violence (*ahiṁsā*), the central focus of Buddhist and Jain ethics. The aspect

> was less important in other religious sects. Veiled, ambiguous references can be culled from the Upanishads, but the exposition of the idea as an ethical value was that of Mahavira and the Buddha. The Jaina understanding of *ahimsa* appears to be an extreme position involving all created beings and attempt to preserve them from destruction, whether deliberate or accidental. The Buddhists tend to stress the ethical question of man's actions in furthering or preventing violence.
>
> — Romila Thapar, 1975, p. 127

In Buddhism, there had been bitter criticism of the belief that merit or man's happiness can be achieved by means of animal sacrifice, such as *aśvasmedha*, *puruṣamedha* or *vājapeya*, essential to the Vedic Brāhmaṇical religion. It is largely due to Buddhist and Jain disputation that the prohibition was extended from cattle to violence *per se*.

> *Ahimsa* can also be explained as a reaction among the new urban groups to the prestige economy of non-urban societies, who were wilfully destroying wealth to no purpose. Sacrifice, it was argued, is essentially an offering;

> consequently it lies not in the destruction of life but in the embodiment of moral values that become the foundation for ethical behaviour — in honouring parents, in honouring all the members of the household from the highest to the lowest, in having patience, meekness, and self-control.
>
> — Thapar, p. 127

Strong similarities in form and content are evident in these respects between the Buddhist precepts and the edicts and codifications in *Arthaśāstra*. These aspects, that is the injunctions related to desirable human attitudes towards the organic world, towards self, family, neighbour, natural elements, resources, etc., have been discussed earlier of this work. One has only to recollect those to grasp the similarities indicated.

It will be utterly naive to attribute to Buddhism-Jainism or other non-Brāhamaṇical heterodox sects, all the ideas related to care for nature, preservation of biosphere, protection of flora and fauna, etc., contained in *Arthaśāstra*. *Athaśāstra* with strong tilt towards Brāhmaṇism and Vedic religion and rituals (vide chapters 2 and 3 of Book 1[1], 3.20. 16[2] and many other *sūtras* or rituals, propitiations, deity-worships, etc., spread through different *adhikaraṇas*) begins with salutation to Vedic teachers and records that this treatise is a compendium of works of all the ancient teachers on science and politics for acquisition and protection of the earth. These ancient literatures like Veda, *Manu- saṁhitā* (supposed to be around 1000 BC) and others predating *Arthaśāstra* contains references to physical environment, mainly based on the basic concept that five primary elements of physical environment, that is land, water, fire, air and space are ever vibrant and charged with vital force. All these basic elements were held with great veneration as being living and sacred entities, and are to be considered as absolutely inviolable. And any form of violation, like for instance cutting of fruit trees, shrubs,

---

1. Chapter 2 and 3 of Book 1 deal with establishing (the necessity of) philosophy and establishing (the necessity of) the Vedic Lore respectively.
2. *Sūtra* 3.20.16 indicates the hostile attitude towards brāhmaṇa religion and sects like Śākayas and Ājīvikas.

creepers or flowering plants, a person for his atonement should be punished by making him mutter the prescribed Vedic hymns one hundred times (*Manu- smṛti,* 11/142). *Manu-smṛti* contains such topics as water management, water pollution, conservation of biodiversity, waste disposal, etc. (Dasgupta, 1995, pp. 29-39). Essence of the ancient enunciations is to be found in the *sūtra*s of *Arthaśāstra,* but in more matter-of-fact form. The mundanity of the codifications, different from precepts of *Arthaśāstra* bear the imprint of Buddha's teaching of adopting practical methods in way of life. Buddhist social code, stressed broad ethical principles of general application to a variety of social groups. Buddhism was more analytical than earlier thinkers in their views on man and society (Thapar, *op. cit.,* p. 130), and certainly on physical environment.

The perceptions about the world around man, and on the human attitudes towards nature, as emerge from the *sūtra*s in *Arthaśāstra,* are certainly contrary to the concept of 'anthropocentric dualism of being and having, namely, to be human and to have everything else at our disposal', the concept which explains the present state of degradation of environment, consequent upon the uneven economic development of several centuries, and callous human attitudes. The concept of 'co-natural word' which takes into account man's relation to surroundings as an integral part and not differentiated separate entities, is implicit in a number of *sūtra*s.

There had certainly been a breakthrough in the conceptual framework as evident in *Arthaśāstra* — towards a new paradigm of mundanity, an outcome of assimilation and amalgamation of earlier concepts with emerging ones, and being much different from ritualisticstringencies. This embryonic philosophy is observed in full bloom in the present-day concept and concern constantly being proferred against man's profligate attitudes towards nature and environment. Nothing else, perhaps, is required to look for in order to gauge the relevance of *Arthaśāstra* even today, more than two thousand years after it was compiled.

6

On Antiquity of Arthaśāstra

NO SOONER the manuscript of *Arthaśātra* by Kauṭilya was brought to light in earlier part of this century, than it attracted the attention of the intellectual world. The reason was many sided. With its discovery, this treatise established itself as the 'earliest and most important textbook specifically devoted to state craft' (Basham, 1954, p. 79). The knowledge about the existence of a highly civilised, advanced administrative, scientific and technological concept in ancient India, was radically altered with the discovery of this treatise, attributed to Kauṭilya who belonged to the fourth century BC. The treatise describes logical, cogent and well-defined rules and regulations, duties and responsibilities of the king and his subordinates for the effective administration of a country. These definitions are backed by the underlying philosophies, also enumerated in the text in unambiguous forms, illustrating the logic and axioms used in the analysis and formulation of the rules, regulations and codes (Kangle, 1986, pt. II, pp. 512-16). These are more obvious in the spheres of administration, encompassing various aspects of life of rural and urban societies.

To attribute that the knowledge contained in *Arthaśāstra*, so advanced even by modern standards, originated in the hoary past is predicated to the acceptance of its being wrtten in the fourth century BC by Kauṭilya who was reputed to be the most shrewd politician of that period and who subsequently became the Prime Minister of the Mauryan kingdom. This fact has not been universally accepted by all the authorities.

Volumes have been written on this issue (Kane, 1968, pp. 149-

256) ever since the exposition of *Arthaśāstra* by Dr. R. Shamasastry (1908). Only a few of these are being quoted here only to reveal the nature of analysis on the date of the treatise. The points on which the theory that *Arthaśāstra* belonged to the fourth century BC were based can be classed in the form of the following three views (Roy, 1925, pp. 171-5, 201-5):

1. The work might have originated with Kauṭilya, but was developed and brought to its present condition by his school.
2. The work was itself originated and developed by a school of polity which was associated in latter times with the name.
3. The work might itself have been composed by one single author or at least one compiler or editor around the third century AD and been fathered on the legendary Cāṇakya Kauṭilya, who was then looked upon as the archetype of cunning and unscrupulous minister.

Amongst those who held that *Arthaśāstra* cannot be relegated to fourth century BC but should be ascribed to much later date, the more familiar authorities are Hillebrandt, Jolly, Keith, Winternitz, Basham, Nag and others. Jacobi opposes their views with various arguments.

Hillebrandt says

> We cannot assume Kauṭilya himself is the sole author of the text in question. It only originates in his school, which quotes often the opinions of other teachers and opposes to them (after the manner of *sūtra* works) expressly the view of Kauṭilya himself, at times expressing the latter in the form of direct maxims.

Hermann Jacobi examines this point and counters the argument by Hillebrandt as follows (Jacobi, 1918, pp. 157-61, 187-95):

> The occurrence of the expression *iti* Kauṭilya, as far as I know, the only argument that has been advanced against the authorship of Kauṭilya . . . when an author, after refuting the views of his opponents, wishes to express his own

> views, he must either speak of himself in the first person, or specify his name. The use of the first person, which involves the bringing into prominence of one's own person, is opposed even this day to the sentiment of Indian scholars; they rather would take pains to conceal their personality. Consequently those authors could not help giving their own views. For this reason it is not right to assert that our *Arthasastra* was written not by Kautilya himself but by some-one from the circle of his pupils, notwithstanding the frequent repetition of the formula *iti Kautilya* in the work.

Jacobi's extended arguments answer all the views, held by different authorities as classified above. It is, therefore, considered appropriate to reproduce further relevant extracts from Jacobi. He states,

> when we say that a work had its origin in the school of a certain individual person after whom it is named, we must assume two things:
>
> 1. That the alleged author was the founder of a school that acknowledged him as such in the sense that he, either materially or formally, brought the development of a 'discipline' to a certain completion and, through its being regularly handed down from teacher to pupil, made a new beginning,
>
> 2. That the 'discipline', that had been handed down in that manner and perhaps also simplified as regards minor details through discussion and controversy, was put forward in the form of a book by some subsequent adherent of the school.

After examining the above questions, it was concluded that

> . . . If one may speak of a school of Kauṭilya in any sense of the term whatever, then such a school could have originated not with Kautilya personally, but only through the medium of the *Arthasastra* written by him. In other words the book does not owe its existence to the school, but the school to the

> book. It is perhaps not superfluous to point out that the word school is used in the last sentence in two widely different senses. In the former case — that is, had Kautilya himself founded the school — the word school signifies the sequence of teachers and pupils, *guru sisyasamtana*, in the latter the totality of the followers of his doctrines, *tanmatanusarita*.

Further conclusions on the subject have been put forth by Jacobi as,

> From the data of Kautilya we can infer regarding the development of the *Arthasastra* that it was at first cultivated and handed down in schools and that subsequently individual authors wrote on the subject. The evolution was already completed before the time of Kautilya, whose work bears the stamps of a strong individuality, both as regards the form and the contents. . . . I do not wish to dispute that Kautilya may have had collaborators for certain parts of his work, especially for such as deal with technical details; officers engaged in the respective branches of administration may have supplied the material and he may have only attended to its editing. . . .

He finally opines that,

> The outcome of our investigation is, on the one hand, that the suspicion against the authenticity of the Kautilya is unfounded, and on the other, that the unanimous Indian tradition according to which Kautilya is the work of the famous minister of Candragupta, is most emphatically confirmed through a series of internal proofs.

Yet doubts have been expressed on other gounds about the authenticity of the work.

> Some authorities still maintain the full authenticity of the work, but there are grave objections to this view.
>
> — Basham *op. cit.*, p. 79

About the points on which Basham doubts the authenticity, he puts forth,

> The text refers to people and places (notably China) which do not seem to have been known to the Indians in the fourth century BC. It does not use much of the official terminology employed in the Asokan inscriptions or in the Pali scriptures, but it contains many governmental terms which apparently did not become popular until post-Mauryan times.

Nag's objections are also on similar arguments. He says,

> Every serious historian will hesitate to consider as written in the fourth century BC a treatise containing names like Harahura and Kapisa, Kamboja and Aratta, Bahlika and Vanayu (Arabia), Tamraparni and Pandyakavataka, Suvarnakundya and Suvarnabhumi, Cina and Nepala.
>
> — Nag, 1923, quoted by Kangle, pt. III, p. 74

In the first point of objection, as noted above, the Indo-China contact has been attributed to a time much after the fourth century BC. Basham further maintians that

> there is no archaeological evidence of cities of the Harappa people beyond the Indus basin.
>
> — Basham, *op. cit.*, p. 25

Recent archaeological findings definitely provide a different picture.

The excavation at Daojali Hading in North Cachar Hills brought out rich deposit including large number of stone axes, bones, and quantities of sherds of chord-impressed and striated beater-impressed pottery. Among the stone tools were large numbers of small ground axes of rounded form, and a number of small shouldered adzes with angular outline. . . .

> They are types which have an extensive distribution in China and south-east Asia and a long and respectable ancestry there. So too has the pottery, in so far as it can be identified. Its affinities seem to be with China, recalling the

pottery of the Neolithic sites around Hongkong.

— Allchin *et al.*, 1968, p. 170

The archaeological findings east of the Indus basin have been summarised as below:

> We find evidence in at least three regions of pre-metal Neolothic cultures, albeit in absolute time they may be somewhat later than Neolithic stages farther west. At least the first two of these are now dated by runs of radio carbon to what in absolute time is a pre-Harappan period. In Kshmir the culture is probably associated with a hunting economy, and there is as yet no proof of agriculture or domestication. Equipment and burial customs lead us to associate the Neolithic of Kashmir with Neolithic or near-Neolithic cultures right across central Asia into southern Siberia and the north-western frontiers of China. . . . In Assam there is a suggestion of an 'Early' Neolithic with possible relations with the two previous groups, and also perhaps with those of south China, and there is more certain evidence of a mature Neolithic with very definite south Chinese and south-east Asia affinities. — Allchin, *op. cit.*, pp. 177-8

As regards the existence of Harappan people beyond the Indus basin, the recent excavations in the Dāmodar river valley in West Bengal and in areas around, have set at rest any doubt about the Harappan, even the pre-Harappan civilisation spreading as far east as Assam, through West Bengal. Excavations in Pāṇḍu Rājār Dhibi in the Burdwān district within Dāmodar river valley have established the presence of Harappan chalcolithic culture in those areas, proving its radio-carbon date as far back as 1012 ± 120 BC; at Mahiṣadal in the adjacent Midnāpur district, the C_{14} dates of 1380 and 850 BC have been assigned to period I and 690 BC to the period II (Allchin, pp. 198-9). More recent findings at Chidahati in the Bhāgīrathī river valley within the Burdwān district in West Bengal, has established Megalithic to Iron age civilisation in these parts of India, compared to Mahenjo-Dāro, Harappā and Ahār (in Rajasthan) civilisations.

Authorities reconstructing the pre-historic and proto-historic human civilisation have resorted to the typological characteristics of artefacts in different regions, materials, rocks, metals, etc., in use, to establish the antiquity and also affinities with different centres of civilisation. Minerals and ores occurring in nature have been taken into consideration with regard to their known geographical occurrences to ascertain whether these were indigenous or imported. In case of artefacts, metals, etc., found in Mahenjo-Dāro, chemical analysis of metals was used to trace the source of some of these, based on the presence of certain characteristic trace elements in them. Thus, the presence of nickel and sulphide in copper implements led to infer that copper ores were, in all probability, obtained from natural sources in Rajasthan, where these are characterised by sulphur compound of copper and nickel, compared to the absence of nickel in copper ores of nearby Western countries like Baluchistan, Iran, etc. Evidences from ore deposits all over the world reveal their strikingly non-uniform distribution patterns which can vary according to region of their occurrence, time of formation, type of metal content, etc. — all of which are related to mineralisation, geologic and 'seemingly unrelated geologic phenomena' (Mookherjee, 1992). It is possible, by modern analysis, to distinguish and even to assign regional source of ores of metals.

Some findings in Mahenjo-Dāro are significant. A large number of minerals, ores and rocks were known and were in use among the Indus Valley people. Lapis lazuli, turquoise, onyx, bitumen, rock crystal, limestone, soapstone, amethyst, sodalite, Jade, lollingite, jaspar, etc., have been found mostly in the form of beads and pendants for ornamental purposes. Cerussite and cinnabar (ore of mercury) have also been found there. They were probably used for cosmetic and medicinal purposes (Roy 1956). From the knowledge of their geological occurrences, it can definitely be concluded that a number of such minerals and ores were brought in Mahenjo-Dāro from outside. Cinnabar stands out, in this respect, as a marker material, since its occurrence in Asia is known from nearby country of China only. Other known occurrences in the classical world were furthest in Spain and Rome.

Thus, archaeological evidence establishes that cultural connection did exist between India and China ever since the dawn of human civilisation.

The great battle of Kurukṣetra in *Mahābhārata* (whose age according to Basham should be the ninth century BC), is interpreted as the conquest of the Kurus by a tribe of Mongol type from the hills (Basham, *op. cit.*, p. 39). And in the hills and beyond, the most advanced civilisation since 2000 BC has been known, amongst Mongoloid races, in China. *Mahābhārata*, which according to Basham should belong to ninth century BC (and not earlier, as some authorities believe) contains references of people from China. Two excerpts are significant

> Pragjyotisa was surrounded by mountain men and Chinese and many other wariors who live close to the ocean beaches.

Further,

> Chinese men I saw and Huns, Scythians, Orisscans, and the ones who live in the interior forest, Varsneyas, Robber Huns, Black Folk and Himalayans — I cannot get over the many who came and were denied admission, yet paid tribute to him, of so many shapes and numerous. . . .
>
> (Tr. by Van Buitenen, 1975, pp. 8 and 117, of 2.23.19 and 2.47.19 of the critical edition of *Mahābhārata* ed. by Bhandarkar and others, Poona).

India's contact with China earlier to the time of Kauṭilya in the fourth century BC cannot be ruled out completely, as assumed by Basham. And hence, the conclusion based on this concept that *Arthaśāstra* belongs to much later date than fourth century BC, is perhaps not tenable.

Studies on the migration of Mongoloid people from beyond the Himalāyas into India, made on the basis of clues from ancient Sanskrit literatures are also interesting.

> It is the consensus of opinion among Indologists that in Sanskrit the terms Kirāta indicated the wild non-Aryan tribes living in the mountains, particularly the Himalayas and in the north-eastern areas of India, who are Mongoloid in origin. These Kiratas were connected with the Cinas or the Chinese, the Bhotas or the Tibetans, and the other Mongoloid peoples. . . . — Chatterjee, 1974; p. 26

These Kirātas have been mentioned in the Vedas.

> The four books of the Vedas were compiled in the likelihood in the tenth century BC, so that the passages in *Yajurveda* and the *Atharvaveda* mentioning the Kiratas are at least as old as that period. When the *Mahabharata* and *Ramayana* were taking shape, between 500 BC to AD 400, particularly in the pre-Christian centuries, they had occupied the southern tracts of the Himalayas and the whole of north-eastern India, North Bihar contiguous to Nepal and to the north of the Ganges, the greater part of Bengal, and Assam including the areas through which the Ganges (the Padma or Padda of the present day) passed into the sea. . . . A meagre stream of trade from China used to filter through this Kirāta country into the parts of Gangetic India. . . . Chang K'ien, the Chinese general and explorer of Central Asia in the second century BC, also testified that there was a trade between north-eastern India and south-western China through routes among the southern barbarian peoples which were not officially known to the Chinese ruling classes in the Capital City in north China; this trade was in the Chinese silk cloth and Chinese bamboo-flutes, among other things, and it brought these articles into Eastern India, which were then carried through the entire north India to as far west as Afghanistan and Central Asia. The kinsmen of the Chinese, the Indian Mongoloids, or Indo-Mongoloids, were the intermediaries in this trade, and not only did they carry material goods from China, but also at times brought ideas. . . .

Regarding their time of migration to India, it is postulated that

> ... They entered probably through Assam, and their advent in the east might have been as old as that of the Aryans in the west, at some period before 1000 BC. By that time they might have pushed along the Himalayan slopes as far west as the Eastern Punjab Hills. They came to be known to the Vedic Aryans as a cave dwelling people from whom the Aryans obtained mountain produce like drugs and herbs and the Soma plant. ... — Chatterjee, 1974, pp. 36-7

It is relevant to quote another opinion, which states,

> And in this connection it is to be observed that the reference to China, instead of proving post-Maurya character of *Arthasastra*, may demonstrate just the opposite. For, China as the name of a 'power' began to have an international reputation with Tsin Shi Hwangti, the contemporary of Asoka. — Sarkar, 1928, p. 382

Thus, it may not be correct to assume that Indians had no knowledge of China during or before the fourth century BC, and the proposition that the antiquity of *Arthaśāstra* should accordingly be considered as later than that time, may not be justified.

Jolly (1923; pp. 42-3) pushes forward age of *Arthaśāstra* to the third century AD, based on expressions of alchemy, particularly in gold making from the chapter on Survarṇādhyakṣa (II.13). The *sūtras* which are taken as processes in alchemy, are interpretable in altogether different forms, based on the etymology of some of the key Sanskrit words and reading them cogently together (Haldar and Sen Sarma, 1994). This has been attempted in the present volume in the chapter dealing with mines and minerals. What has been interpreted as conversion of base metals into gold by amalgamation with mercury (obtained from Cinnabar), is actually a process of winning and separating gold from ores containing this metal in combination with silver or other metals. This is not alchemy, *per se*, but an accepted process even in modern metallurgy.

As a matter of fact the Indian contribution to the development of the science of geochemistry, was never studied or considered seriously. As late as 1984, an exponent on the subject, while dealing with the pre-scientific roots — philosophy and alchemy from antiquity till the end of the sixteenth century AD — set forth the premise as

> Early Chinese and Indian speculations, however interesting they might be, are disregarded here, because they seem to have had no direct impact on the development of geochemistry. — Dudich, 1984, pp. 1-16

His emphasis was on Aristotle's concept of four elements — earth, water, air and fire, and their interactions which are responsible for the origin of rocks, metals, etc., as the embryonic perception, from which he proceeds to trace the history of geochemistry and cosmochemistry. In this attempt, the fact was disregarded that *kṣiti* (earth), *ap* (water), *teja* (heat), *marut* (wind) and *vyoma* (space), considered in Indian philosophy as the basic elements in cosmology, were dealt with in earliest Sanskrit literatures of Vedas. Without such consideration, evaluation of the type expressed in the above excerpt amounts to peremptory dispensation and disinformation. Fersman, considered to be the father of modern geochemistry truly said that

> The history of geochemistry has not been written so far. Even the materials for it have not been collected yet'.
>
> — quoted by Dudich

Thus, there had been various attempts to question the date and authenticity of *Arthaśāstra*. As has been revealed from the above brief discussions, it is apparent that such attempts are based on very slender and simplistic assumptions, which really do not hold on at close scrutiny.

At the same time questions had been raised about the autochthonism of *Arthaśāstra*, or in other words, some authorities believed that knowledge in different spheres, were based on ideas borrowed from Western civilisation. Some even hold such drastic opinion as, speaking about alchemy in India,

> Regarding the age and origin of Indian Alchemy, the opinion of those does not deserve much credit who believe it to have been the outcome of Indian intellectual activity (*loc. cit.* Roy, P.C.).
>
> —Jolly, *op. cit.*

Though written in the sixteenth century AD, that is about 1800 years after Kauṭilya, and having startling similarity with the contents of *Arthaśāstra*, especially the philosophy of the royal polity, the book *The Prince* by Niccolo Machiavelli has not been considered anywhere to have been influenced by *Arthaśāstra*. Kauṭilya has been described by several authorities as the Indian Machiavelli rather than the other way round. This well-known and much publicised comparison is based mainly on the political contents of the treatise. The attitude originates from the assumptions that anything relating to science and technology had their origin in Greece and Europe. This Euro-centric idea is displayed in the wiritings of many scholars of the West and is often implicit in many scholars of the East, may be in a milder way, in attributing benefit of doubt to the originality of Indian scientific principles and technological innovations. On these points a valuable synoptic picture has been depicted (Chattopadhyaya, 1986, chapters on preface and introduction). Some of the points are worth quoting.

> The story, we are told, began with some kind of Greek 'miracle' resulting in the dawn of science. Not that the facts of the older civilisations — of Egypt and Mesopotamia, and, in recent years, also of the Indus — are denied. But these are mentioned cursorily and mainly for the purpose of showing why there could be no real science before the Greeks.

Chattopadhyaya quotes Arnold Reymond (from his book *Science in Greco-Roman Antiquity*) as

> compared to the empirical and fragmentary knowledge, which the peoples of the East had laboriously gathered together during long centuries, Greek science constitutes a veritable miracle.

Against the idea held by Paul Tannery that Indians borrowed the use of 'Sines' in trigonometry from the Greeks, there are contrary opinions in favour of the fact that Indians knew of Sines. J. Filliozat mentions that

> trigonometric sine was not mentioned by Greek mathematicians and astronomers, that it was used in India from the Gupta period onwards . . . that the Arab astronomers knew them from their Indian contacts and passed them on to Europe in the twelfth century AD, when the work of al-Battani was translated into Latin.

Filliozat laments that

> If this is the way we are to argue, there was never any science other than Greek science, and the question whether science has any origin other than the Greek 'miracle' is solved in advance. —quoted in Chattapadhyaya, p. 44

J.D. Bernal, on the strength of the results reached in the great work of Joseph Needham on Science and Civilisation in China, commented (quoted from his *Science in History* by Chattapadhyaya, pp. 44) —

> The technical advances of the Middle Ages were made possible by the exploitation and development of inventions and discoveries which, taken together, were to give Europeans greater powers of controlling and ultimately of understanding the world than they could get from the classical heritage. Significantly, the major inventions . . . were not themselves developed in feudal Europe. All seem to have come from the East, and most of them ultimately from China. . . . Already enough is known to show that the whole concept of the superiority of Western Christain Civilisation is one based on an arrogant ignorance of the rest of the world.

The Euro-centric egoism is again an outcome of several deeper motivations, mainly the desire for social, religious and political domination.

Joseph Needham does not mince his words on these points

> .... if ... you wish to explain why Europeans were able to do what Chinese and Indian were not, then you are driven back upon an inescapable dilemma. One of its horns is called pure chance, the other is racialism however disguised. To attribute the origin of modern science entirely to chance is to declare the bankruptcy of history as a form of enlightenment of human mind. To dwell upon geography and harp upon climate as chance factors will not save the situation, for it brings you straight into the question of city-states, maritime commerce, agriculture and the like, concrete factors with which, autonomism declines to have anything to do. The 'Greek miracle', like the scientific revolution itself, is then doomed to remain miraculous. But what is alternative to chance? Only the doctrine that one particular group of peoples, in this case the European 'race', possessed some intrinsic superiority to all other groups of peoples. Against the scientific study of human races, physical anthropology, comparative haematology, and the like, there can of course be no objection, but the doctrine of European superiority is racialism in the polititcal sense and has nothing in common with science.
>
> — quoted by Chattopadhyaya, *ibid.*, p. 53

As for the work on reconstruction of the origin and development of science and scientific thought in Indian history, as used to be practised, Chattopadhyaya (*ibid.*, pp. 46-7) recounts:

> The view of science being a monopoly of Western Europe has other undersirable consequences. It serves the forces of conservatism within the Asian countries, which in defence of stagnation and status quo, prefer to cut off Asian culture from the mainstream of global science. This makes it covenient to project the irrational religious-mystical trends of the past as representing the quintessence of Asian culture.

Perhaps this influenced many, at least in their subconscious thought process when dealing with the history of science in ancient India, as evident from the confusing assessments given about the development of ancient scientific thoughts in India. We may take an instance. Sepaking about Chemistry in India, it has been stated that

> Chemistry in India was developed empirically and occupied itself, more or less, with collection of *accidentally* discovered facts associated with various practical arts like ceramics, metallurgy, metal-working, and medicinal preparations without any recognition of the *chemical principles or nature of the chemical changes* involved in their pursuit. The result was that the thoughts and ideas could not germinate *into scientific laws and theories based on experimental observations* and verifications. Likewise, the mechanical skill displayed in the pursuit of practical arts could not develop into technology in the absence of guidance and suggestions from scientific knowledge. *Chemistry was dominated more by seeing and believing than by thinking and knowing.* . . .
>
> Nevertheless India's achievements in the use of minerals, metallurgical techniques, processing of chemicals of everyday use, extraction of metals from their ores, and craftsmanship in the manufacture of certain metal products, *which required mastery of some chemical process*, were quite remarkable. Some of the technical skills exhibited by ancient Indian chemists and metallurgists were indeed noteworthy.
>
> — Priyada Ranjan Roy, 1986, pp. 150-1 (emphasis added)

It needs no emphasis to demonstrate the contradictory nature of the assessments. But since the assessments are based on certain terms like 'empirical', 'law', 'theory' — used in scientific phraseology, it demands somewhat elaboration, to establish that the terms, which have been used in the above passage to undersize, albeit faltering tribute to the ancient thoughts, developments and achievements in scientific line, are their real strength and solid foundation, on which subsequent developments took place in the real sense of modern science.

It is indeed quite appropriate to indicate the place of empirical concepts and ideas in the evolution of such knowledge, known as scientific, and related terms of 'theory', 'hypothesis', 'law', etc.

> Theories are imaginative constructions of human mind — the results of philosophical and aesthetic judgements as well as of observation — for they are only suggested by observational information rather than inductively generalised from it. Moreover, theories cannot ordinarily be tested and accepted on the same grounds as laws. Thus, whereas an empirical law expresses a unifying relationship among a small selection of observables, scientific theories have much greater scope, explaining a variety of such laws and predicting others as yet undiscovered.
>
> — *Encyclopaedia-Britannica*, 1985

Going deeper into the meaning and methods, it may be observed that

> There is, in fact, nothing to which we can apply rules of method, because the laws of nature must be in our possession before we can treat them. If there were any rule of inductive method, it would direct us to make an exhaustive arrangement of facts in all possible orders. . . . It is the method by which the first simple steps are taken in an incipient branch of science.
>
> Before the dignified name of science is applicable, some coincidences will force themselves upon the attention. Before there was a science of meteorology observant persons learned to associate clearness of atmosphere with coming rain, and a colourless sunset with fine weather. Knowledge of this kind is called empirical, as seeming to come directly from experience; and there is considerable portion of knowledge which bears this character.
>
> — W. Stanley Jevons, pp. 505-6

In the background of the knowledge of different branches of science at the time of Jevons (towards the end of the last century), he states that

> We must not indeed suppose that any science will ever entirely cease to be empirical. Multitudes of phenomena have been explained by the undulatory theory of light, but there yet remain many facts to be treated. The natural colours of bodies and the rays given off by them when heated, are unexplained, and yield few empirical coincidences . . . even the establishment of a wide and true law of nature is but the starting point for the discovery of exceptions and divergences giving a new scope to empirical discovery.
>
> — Jevons, *op. cit.*, pp. 528-9

The state of science of physics of today, much different from what has been stated above, truly depicts the extensive form of human knowledge consisting of facts which have been observed empirically and which have subsequently been brought in accordance with other facts by hypothesis concerning the general laws applying to them. This is the knowledge which may be said to have been explained, reasoned and generalised. Since

> The process of science are characteristic of human action in that they move by union of any empirical fact and rational thought, in a way which cannot be entangled. There is in science, as in all our lives, a continuous to and fro of actual discovery, then of thought about the implications of what we have discovered, and so back to the fact for testing and discovery — a step by step of experiment and theory, left, right, left, right for ever. — Bronowski J., p. 36

Thus, to relegate the ideas as inconsequential in scientific development because of their being empirical, is itself unscientific — since the assessment overlooks the very philosophy of evolution of human thoughts. The best way to understand the ancient thoughts and reconstructing their history (a subject by itself, complex and vast), it is imperative to define clearly the objectives and methods, in the absence of which any assessment made becomes either emotive and illogical or diametrically opposite attitude of derision of anything that belongs to the hoary past.

We believe, with adequate evidence and reasons behind, that fourth century BC was an illuminating landmark in Indian history, with apodictical clarity and obviousness in transcendental thoughts and ideas. Karl Jasper's identification of the characteristics of the age of transcendence has been taken as the guiding principle in the present study, analysing the observations and codes documented in the *Arthaśāstra* in the light of the modern science. Jasper says of the 'age of transcendence' as

> It was then that our intellectual, moral and religious civilisation was born and that the foundations were laid on which we continue to build, despite difference in the superstructures we have erected and go on erecting.
>
> — Karl Jasper, quoted by Eric Weil, 1975

The question of autochthonism and originality of the thoughts and ideas contained in *Arthaśāstra* crops up next as the most important one, particularly in the face of the concept of Euro-centric science discussed earlier. In this respect, Filliozat (quoted in Chattophadhyaya, pp. 44-5) says that

> only a profound study of Indian scientific developments in parallel with those which took place elsewhere about the same times can reveal the degree of originality of that science, and hence enable us to understand the role which India played in the history of the growth of man's knowledge of nature.

We may examine this issue.

The most widely covered scientific field in *Arthaśāstra* relates to minerals and ores, forming part of geological science, and mining, also related to geology. A search for the beginning of geological science in Europe may commence with a look into the earliest literature of the continent. These are of the early Greeks, since the rest of Europe has not left any records of similar interest or achievement. The early 'sophi' or 'wise men of Greece', who later came to be known as 'philosophers' were engaged in many daring speculations concerning the origin of the universe (Adams, 1938, pp. 8-9).

Referenc es could be obtained of a number of Greek philosophers who have contributed something or other to geological science, dating back from seventh century BC, starting with Thales of Miletus (636-546 BC), up to the time of Kauṭilya in fourth century BC. Other names which figure prominently as having contributed some ideas to the science of geology are Anaximander (610-547 BC), Pythagoras (540-510 BC), Xenophanes (540-510 BC), Herodotus (around 484 BC), Empedoclese (around 444 BC), Democritus (460-357 BC), Aristotle (384-322 BC), and Theophrastus (370-287 BC). The references scattered through the mass of literature, revived by Julius Schwarcz in 1868, do not present formal treatment of any geological question, but reference to geological phenomena are rather made *en passant*. Of the early Greek writers, Aristotle and Theophrastus stand out prominently. In *Meteorologica* Aristotle recognises two types of 'exhalations' which he considers, as responsible for various atmospheric and geological phenomena. His 'moist exhalation', of the nature of vapour, originating from the body of water due to sun's heating, condenses to cloud. The second kind of exhalation, which he terms 'dry exhalation', and which is like smoke, under certain circumstances causes thunder, lightning and other atmospheric phenomena, and at other times it gathers as thick dark cloud which condenses to form stones, and when cast down upon the earth with great violence, are called 'thunder bolts'. He states that

> just as their two-fold nature gives rise to varieties effects in the upper region, so here it develops two varieties of bodies that originate in the earth, namely fossils and metals.

The terms fossil is used for anything dug out of the earth, and Aristotle makes special use of the term to include

> such kinds of stones which cannot be melted, together with realgar, ochre, sulphur and other things of that kind. Metals are distinguished from these as those bodies which are either fusible or malleable, such as iron, copper and gold.

In summary, Aristotle thought that

> stones grew on the surface of the earth or in the earth's

> interior through the influence of the heat or the rays given off by the sun or other heavenly bodies, penetrating into earth's crust and there in certain places where the conditions were favourable, causing exhalations, which bringing about new combinations of elements, give rise to stones of various kinds. — Adams, *op. cit.*, p. 81

Theophrastus is considered to be more matter-of-fact in dealing with minerals, in his work entitled *Concerning Stones*. The minerals, in his treatment, has been classified into two categories, namely, those which can be changed or melted in some manner by heat, and those which remain unaltered when submitted to the action of fire. One aspect of his treatment which stands out as unique amongst the classical philosophers dealing with minerals, is that he continually refers to the use of minerals to which they may be put and to their various applications in the arts and industries.

He refers to the principal gems in terms of males and females, mentions about the attracting properties of lodestone, about pearls growing in the sea, about varieties of marbles, alabaster, gypsum, ochre, and considers also a few metallic ores. There are many, amongst them Anaxagoras (500-428 BC) and Theophrastus (370-287 BC) who hold the view that the metals and metallic ores, having once come into existence, in whatever manner it may be, are endowed with the power of growth and propagation. Theophrastus refers to metallic veins in the same term. The same idea continued for subsequent 1800 years or even more. Agricola, known as the father of mineralogy, held the similar idea very forcefully as late as the sixteenth century AD. After exhaustive studies of the Greek and classical literatures available of that period Julius Schwarcz opines that

> . . . while some of the Greeks in these early times observed certain striking geological phenomena and speculated in a general way with reference to their cause and origin, there have come down to us from the Greeks but very few contributions of any importance of geological science.
>
> — Adams, *op. cit.*, p. 10

Adams himself comments,

> It may be noted that the ancient writers usually made no distinctions between gems, common minerals and rocks. The diamond and millstone appear together in their lists of 'lepides' or stones. Nothing was known in those ancient times concurring either *the chemical composition or crystallographic form of minerals, although these are now considered to be the most important factors in the distinction of mineral species.*
>
> — Adams *op. cit.*, p. 77 (emphasis added)

He quotes another authority of the sixteenth century,

> Medieval mineralogy in fact was not a science . . . but a fairy castle, the insubstantial fabric of a dream, often quaint and even beautiful, but destined to crumble away because it had no foundation in reality. It was now to be succeeded by a true science of mineralogy built upon the basis of close observation and diligent study of the materials of the earth's crust. — Adams, *op. cit.*, p. 169

The knowledge, ideas and philosophy concerning the geological science, and mineralogy in particular, as recorded in the *Arthaśāstra*, depict very sharply and distinctively the modernness, when compared with the contemporary classical Western literatures on the subject. The very deficiencies for which the ideas have been branded as 'fabric of a dream, often quaint', can be seen to be altogether non-existent in *Arthaśāstra*. On the contrary, the ideas, basic premises and their applications are in such well organised form as rendering the strength to the scienticism of *Arthaśāstra*.

Regarding the quaintness of ideas and their subsequent logical evolution, it is said that

> If then, casting our thoughts back through the long succession of past centuries, we ask why the ancients entertained such quaint and curious views concerning the origin of metals and their ores, we find that these views

> were quaint because they were based almost entirely on *philosophical speculation* and not on *observation*. It was not until the time of Renaissance, when men like Agricola arose, whose views were based on practical knowledge gained by a long continued study of the structure of the earth's crust in the great mining districts of Europe, that the commencement was made of true understanding of the subject. — Adams, p. 328 (emphasis added)

The author of *Arthaśāstra* was not a speculative philosopher by any measure as evident from the subjects dealt with in the treatise, most mundane, practical and matter-of-fact. And it proves beyond doubt his keen observation. In the sphere of gems, minerals and metals — his power of observation can be seen from the way the facts about them have been recorded.

It may be observed that *Arthaśāstra* has attempted, in the field of the science of minerals, metals and ores in —

(1) establishing the clues, in terms of the physical attributes of minerals and ore, for differentiating gems, common minerals and ores.

(2) establishing the identifying clues in plants and water for locating the occurrence of certain types of metals/ores.

(3) establishing the methods of purification and beneficiation of ores, applying various chemical ingredients.

(4) establishing the various forms of economic utilisation of metals, and in particular the alloys of noble metals in combination with other metals.

(5) establishing the methods and clues for identifying the old workings (mining sites), etc.

The book does not contain, by and large, ideas by which it could be assailed of being quaint, baseless and untenable. And lastly, ideas similar to those prevailing in the Western world of that time, as summarised in very brief in the preceding paragraphs, are not to be found in *Arthaśāstra*. It was demonstrated in the chapter dealing with mines and minerals, that viewed in the light of the modern science of minerals and mining, the treatment of these subjects in *Arthaśāstra* startlingly resemble the advanced modern concepts. A.A. Macdonell and H.H. Wilson's perceptions, in addition to those of other authorities discussed earlier, as regards the mutual influences of the ancient Indian and Western ideas and thoughts, are worth recalling. Macdonell in his *History of Sanskrit Literature* remarks:

> According to Greek tradition, Thales, Empedocles, Anaxagoras, Democritus and others undertook journeys to oriental countries in order to study philosophy. Hence, there is at least the historical possibility of the Greeks having been influenced by the Indian thought through Persia.

Prof. Wilson also observes:

> That the Hindus derived any of their philosophical ideas from the Greeks seems very improbable, and if there is any borrowing in the case, the latter were most probably indebted to the former. — Roy, *op. cit.*, pp. 47-8

The facts so far presented, excerptions from authorities recorded and the very purpose of doing such are only with the intention of depicting the manner in which the controversies on the origin of higher ideas and thoughts in science, technology and other spheres of knowledge are articulated. Each of the authorities expressing views in favour of Euro-centrism of the origin of, or in favour of supremacy, of isolated ancient civilisation, may have their respective sources and arguments, but, at least in some periods of civilisation, there exists information to support the 'assumption of an absolute synchronism' which

> would be more meaningful if all the different civilisations could be seen as having lived in a common time, that is to say, as having interacted on each other. Unfortunately, the question of reciprocal influences has fared badly, due mainly, perhaps, to the contrast between the piecemeal character of the borrowings from the one to the other that have been alleged and the very articulate complexity and

frequently self-contained sense of superiority of each.

Against this approach, it has been said that

> We should guard against the assumption that each culture has evolved in isolation: We may be sure that 'commerce' in the widest sense of the term has always gone on, and we should not close the door to hints of interaction.
>
> — Louis Dumont, 1975, p. 153

In spite of such an approach, perhaps very truly suggestive of pragmatic appraisal of the elements of civilisation, there remains the possibility, and which cannot be totally ignored, of the outcome of an altogether different, advanced and illuminating product out of the assimilation of the elements from different streams of civilisation. This may be possible when a society is receptive, and social (political and economic situations included) system is conducive to progressive development. Complexities of social systems, which influence the waxing and waning of the course of progress are available to us in history. The differences, similarities and the modernity of ideas prevalent in the field of geological science, particularly in the branches of mineralogy, gemmology and mining, as contained within the *Arthaśāstra* will be clearly seen in this context when examined in the light of the modern scientific theories.

In the field of Indian art and architecture, a very emphatic account has been provided by Havell, which deserves extended excerption. (Havell, 1915, pp. 2-4). He states:

> The account which Megasthenese has left of the splendour of Chandragupta's imperial capital, Pataliputra, the modern Patna in Eastern Bengal, are of little practical value to the architectural student.

But the archaeological excavation being undertaken at the time when he wrote his book, he maintains that this excavation

> will doubtless produce material of great archaeological interest, but they will never establish any new theory or confirm any old one concerning the origin of Indian art.

These are already so decisively indicated by the great mass of material accumulated in many other places in India that the question whether Asoka or Chandragupta Maurya planned their palaces on Persepolitan models, or built them with foreign craftsmen, becomes entirely a side issue and a matter of local interest only. It will doubtless be used as an argument for British architects in India importing modern Western methods of building — though there is a difference between those which stimulate the craftsman's ideas and those which sterilise them; but it is foolish to argue that Persepolitan fashions at Pataliputra prove that all the great art of India in early Buddhist times was inspired by foreigners — unless all Indians of Aryan race are to be classed as such.

We learn from Megasthenese that the imperial palace was planned on a grand scale like those of Susa and Ekbatana[1], that it was gorgeously decorated with gold and silver and placed in the midst of a fine garden with numerous water ponds — a description which might apply equally well to the palaces of the Great Moguls. It is highly probable that Chandragupta, like his grandson, made considerable use of craftsmen from Persepolis or from Mesopotamia. Indeed, the latest discoveries of Western scholars seem to make it necessary to abandon the theory that the early Aryan invaders of India were only nomads whose practice in the arts was chiefly confined to war and agriculture. If it be true — as the Russian scholar, Sheftelovich asserts that the Kassites, who took Babylon in 1746 BC and established a dynasty there which lasted 600 years, were Aryans speaking Vedic Sanskrit, whose chief god was Surya, Babylon must be regarded as a half-way house of the Aryan race in its march towards the Indus valley, before they entered India, not only the high spiritual culture which is reached in the

---

1. Susa was the main capital and Ekbatana the summer capital of Persia at the time of King Darius I (521-486 BC).

Rig Veda, but a prolonged experience of the civic arts, including architecture.

Recent German excavations on the site of Babylon show that the science of building in Vedic times had advanced much further than has hither to been suspected. The Babylonians had apparently perfected the construction of radiating arches centuries before Rome was founded, and it is probable that both Roman and Indian builders derived their knowledge of it from the same source in Western Asia. When the seat of the imperial power was transferred from Persia to eastern India in the fourth century BC, it must have been as natural an inclination for Chandragupta Maurya to supplement local architectural talent at Pataliputra by bringing experts from Persepolis as it is for Anglo-Indian rulers in the present day to supersede it by experts from London — for Western Asia was the great cultural centre of the Indo-Aryan race.

But it is a great mistake of Anglo-Indian writers to assume that Indian building under the Mauryan dynasty was wholly, or even to a large extent, an importation of 'styles' borrowed from Persepolis.

Chandragupta and Asoka might set Persepolitan fashions in their palaces, just as Indian princes of today try to imitate the fashions of London and Paris; but Indo-Aryan culture had already for many centuries planted itself deep in Indian soil, and Indo-Aryan building had acquired an Indian character as distinctive as that of Indo-Aryan philosophy and religious teaching. . . .

In support of the above contention that architecture had assumed a distinctive character in Indian soil since earlier time, evidence can be had from a number of *sūtras* (Book III) of *Arthaśāstra*. Moreover, Indian architecture during the Mauryan Period can be seen to display originality, in spite of some resemblance with Greek or Persian architecture, when the Indian architectural forms and techniques are closely analysed. This is very well manifest in the Mauryan towers. The Persian columns stand on bell-shaped bases, like the calyx of a flower reversed or on plain rectangular blocks. The column of the rock tomb of Darius at Naksh-i-Rustam has bases with plain circular mouldings. Mauryan pillars are maintined in position by plain slabs of stone or brick work, which are, however, buried beneath the ground, and the pillars have an appearance of standing by their own weight against all the forces acting on them. The concept of imparting stability to a high-rise structure against conceivable forces, by appropriate foundation securely embedded in ground was certainly an addition to engineering technology. (Mitra, 1927, pp. 543-4). Bhattacharya proves that the science of *vāstu-vidyā* was known in times earlier than *Arthaśāstra*; but it was established as a well-organised system in *Arthaśāstra* (Bhattacharya, 1986, p. 11). To put it in other words, the knowledge on *vāstu-vidyā* was more perfected and established almost to mathematical precision during Kauṭilya's time, the proof of which is obtained from the formulations and codifications, almost in the form of, we may be permitted to say, a Manual. Though the details of experimentation, basis of formulation, etc., are not demonstrated in the *Arthaśāstra*, yet the codes of practice in different branches of science and technology dealt in the treatise, are found to be theoretically and practically sound, workable and reproducible in the light of the modern science and technology. It would not be illogical to infer that the codes must have been derived and formulated on adequately sound theoretical foundation and repeated practical application, observation and modifications.

Close examination of the questions of antiquity and autochthonism of *Arthaśāstra*, reduces the debates more or less into a single issue. And that is, those who are prepared to offer any credence to the treatise, are willing to do so only after considering the compiler Kauṭilya and the treatise, as belonging to AD 150 against the other claim of this date being in the fourth century BC. The former group does not believe that Cāṇakya, the king maker and formidable Prime Minister of Candragupta Maurya of the fourth century BC 'did not have a name called Kauṭilya' (Prof. T.R. Trautman, 1968, quoted by Rangarajan, p. 19). The contention of this group is loaded with covert implication that the concepts, ideas, knowledge, practices and prudence could not have attained the level of excellence in India as far back as in the fourth century BC, until imported from and influenced by ideas from the contemporary outside world, much later on.

The opposite claim is based on the historical evidence of Cāṇakya (known as Viṣṇugupta, which was his personal name, 'Cāṇakya' which indicates him to be the son of Cāṇaka and 'Kauṭilya' meaning that he belonged to Kuṭila *gotra*), being the main architect in destroying the Nanda dynasty and installing Candragupta Maurya on the throne of Magadha in 321 BC, immediately after Alexander's invasion of India. It is conceivable that the political turmoil and instability of the country at that time excited an intelligent mind to systematise knowledge on various socio-political and economic affairs available till his time, infusing his acquired experiences into them, in the form of a manual. This has been unequivocally stated by him at the outset.

> This *Arthaśāstra* is a compendium of almost all similar treatises, composed by ancient teachers, on the acquisition and protection of territory. Easy to grasp and understand, free from verbosity, Kauṭilya has composed this treatise with precise words, doctrines and sense.
>
> — *Sūtras* 1.1.1 and 1.1.19; tr. Rangarajan, p. 9ᶜ

The author left no scope for imaginative and extended aʳ . the introduction of contemporary ideas and experiᶜ .ne precepts when he states that

> This treatise has been composed by one who, resulting the misrule of the Nanda kings, rescued this neglected science and used it as a weapon to destroy them and save the kingdom. By following [the principles set out in] this treatise one can not only create and preserve *dharma* [spiritual good], *artha* [material well-being] and *karma* [aesthetic pleasure] but also destroy [their opposites, i.e.] righteousness, material loss and hatred. It is a guide not only for the acquisition of this world but also the next.
>
> — Tr. of *sūtras* 15.1.71-3 by Rangarajan, p. 100

Philosophy and methods followed by Kauṭilya were proved to be archetypal and were followed and reproduced in various periods with variations even in Western world. He generated ideas and philosophies and was himself a creation of his time. This is not just any hollow panegyric of a brilliant mind. There are moments in history which have the characters immanent in them for creative genius. Bronowski ascribes the genius of Newton to such comparable moment in history marked by turmoil and instability.

> These are the moments when the powerful mind and forceful character feeds the ferment of times, when his thoughts quicken, and when he can inject into the uncertainties of others the creative ideas which will strengthen them with power. At such a moment the man who can direct others, in thought and in action, can remake the world.
>
> — Bronowski, *op. cit.*, p. 19

Kauṭilya is exactly the man of this description. Preceding discussions about his ideas, concepts and perception on various subjects bring out his originality.

The question of development of higher order to thoughts, concepts and perceptions amongst man living far apart, without actual contact and communication with one another, has initiated studies, mainly on the nature and evolution of knowledge and psychology from the primitive stage to the most modern man. Various parameters have been considered and the analysis of the complex subject enables us to some workable conclusions. It is believed, as a form of generalisation, that

> similar complexes of material culture arose in similar natural conditions among peoples who were at one level of socio-economic development but were different in origin and often separated by thousands of kilometer of land and ocean. Man's economic activity, of all the phenomena of culture, is linked to the maximum with the geographical environment, so that its trends have been called economic-cultural types in dependence on the connections with the environment

plus the phenomena of culture accompanying economic activity.

— V.P. Alexeev, 1986, pp. 286-7

We have the historical evidence of breakthrough in the philosophy and socio-political set-up in India in the sixth century BC, due to the upshot of 'heterodox sects', of which Buddhism and Jainism had a lasting continuity. It is thought that Buddha, a strict disciplinarian, did not encourage any speculative philosophy and ably developed the rational attitude and humanistic outlook. Buddhist analysis of the source of knowledge, relationship of matter and consciousness, relationship of perception ('a definite and interpreted reaction is known as perception of a particular and special object') and consciousness or awareness, etc., constitutes the most rationalistic approach and certainly makes the step towards evolution of scientific thinking (Haldar, A., 1981, pp. 30 and 72). Profound influence of the Buddhist philosophy on Kauṭilya is transparently evident from proposition, axioms and treatment of various subjects in *Arthaśāstra*.

In the above background and considering the accepted elements in the framework and norms of analysis of the history of epistemology, it can convincingly be inferred that the ideas, concepts, perceptions and practices encapsulated in *Arthaśāstra* reveal transcendental stage in antiquarian evolution, and for which it does not call for revoking any form of externalism.

Note: 'The Persians called for Greek artists to work on grandiose building schemes of Darius and Xerxes at Persepolis, and elsewhere. . . . Thus, long before Alexander's conquests hellenised the entire known world, Greek artists and Greek works of art had reached far afield . . . the chief interest in this penetration lies in its permanent effects upon the arts of the areas it affected. These effects range from poor imitation of imported works to the creative adaptation of Greek ideas into an independent living tradition. It is not possible yet to assess fully all the effects, because of the difficulties in dating much of the materials. It is easy to see similarities, some perhaps fortuitous, between early Indian sculptures and sixth century Kouroi, between seated Buddhas and Ionian figures, between Celto-Lingurian sculptures from provence and later archaic Greek sculpture, but we are still far from understanding the precise relationship between them.' — Donald E. Strong, p. 103 (*Note*: Kouros is the generic term for the standing male statues of the archaic period.)

Appendix 1

Critical Notes on some Engineering Features

Section A: Calculation of the angle of slope of excavation (Sūtra 2. 3.4).

FOLLOWING the *sūtra*, a general diagram can be drawn as shown. The trigonometrical relations of the components of the diagram can be expressed as below:

$\tan A = d/[(x\text{-}y)/2] = 2d/(x\text{-}y)$, when slope angles of both banks are equal.

Therefore $A = \tan^{-1}[2d/(x\text{-}y)]$.

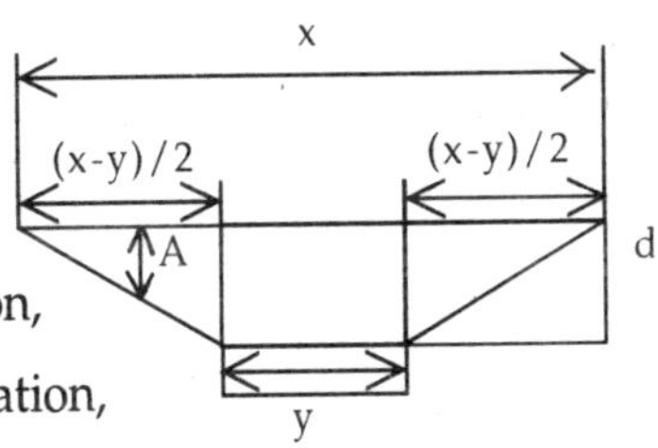

In the diagram

A = slope angle,

x = top width of excavation,

y =bottom width of excavation,

d = depth of excavation.

Various values prescribed in the *sūtra* and corresponding angles determined from the above trigonometrical relations are as below:

Sl.	*x [feet]*	*y [feet]*	*d [feet]*	*x:y:d*	*A [Angle in Degrees]*
1.	84	28	42	1:1/3:1/2	56
2.	84	28	63	1:1/3:3/4	66
3.	72	24	36	1:1/3:1/2	56
4.	72	24	54	1:1/3:3/4	66

5.	60	20	30	1:1/3:1/2	56
6.	60	20	45	1:1/3:3/4	66

It is interesting to note that the dimensions have been fixed in such proportions that the slope angle of excavation would be 56° and 66°, except where the excavation is square (or rectangle) in cross-section, in which case the slope would be vertical or 90°. The dimensions appear to have been given as example only, to highlight the proportionality of top and bottom widths and depths, which would provide these optimum angles on excavation.

Height of slope: An open excavation in a normally consolidated clay soil will stand vertically without any support, provided that the face does not exceed a specific height, called 'critical height', which can be ascertained by knowing two properties of soil, viz., density and cohesion. The former is the soil mass per unit volume (pounds per cubic feet, or kilogram per cubic meter). The latter lends a soil certain degree of stiffness and rigidity. The kind of cohesion is accounted for by rigid bonds between the soil particles.

Cohesion is expressed as pounds per square feet, or in equivalent metric units. The critical height is ascertained as $H_c = [4\,C]/Y$, where C is cohesion and Y is density of soil.

Several values of critical height for various soil consistencies are given below (Tomlinson, 1963, pp. 530-1):

	Very Soft	*Soft*	*Firm*
Cohesion [pounds/sq.feet]	0-375	375-750	750-1000
Y [pounds/cu. feet]	115	125	125
H_c [feet]	13	13-24	24-48

If the height of a slope is greater than H, the slope will not be stable unless the slope angle is less than 90°. Greater the height, smaller must be the angle of slope.

Section B: Design of tower/pillar

Sūtra 2.3.18 states — *stambhasya parikṣepaḥ ṣaḍayamaḥ, dviguṇāḥ nikhataḥ*. The word *ayamaḥ* has been translated as 'length', and

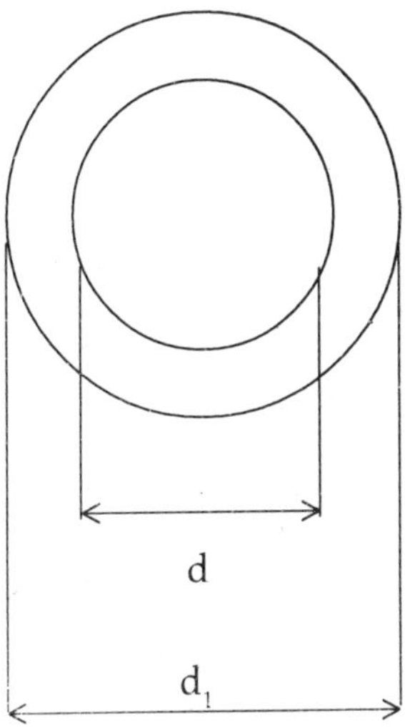

ṣaḍayamaḥ as one-sixth of the length (Kangle, pt. II, p. 65). *Ṣaḍayamaḥ*, like *ṣaḍananaḥ*, should actually mean six times and not one-sixth. Again, considered in terms of mensuration (the branch of mathematics dealing with the dimensional measurements of solid bodies), the word *āyām* also means width or breadth (M.M. Williams-dictionary). We shall consider both the cases for their implications in the design calculations. The dimensional relations as stated in the *sūtra* can be expressed mathematically. In a circular cross-section, the width (*āyām*) obviously means the diameter. The circumference (*parikṣepaḥ*) of a circle is expresed as πd, where d is the diameter and π is a mathematical constant whose value is approximately 3.1416. Thus, the circumference of a circle cannot be six times its diameter. Instead of waiving off the proposition of the *sūtra* as absurd, there is reason to believe that the relationship has been stated with respect to two co-centric circular section, the width having been taken as the diameter of the inner circle. We may consider the diameters of the inner and outer circles as d and d_1 respectively, as drawn in the diagram, d_1 being greater than d. The propostion of the *sūtra* can now be examined on this assumption. The circumference might refer to that of the outer circle, expressed in terms of the diameter of the inner one. This can be put in mathematical form as: $\pi d_1 = 6d$, in which case d_1=1.9d, or say 2d. Thus $d_1 = d + (d/2 + d/2)$.

This stipulation and the mutual relationship of the two co-axial cylindrical structures (the tower actually is cylindrical with circular cross-section) lead us to imagine as inner hollow section, perhaps meant for porviding service facilities, like staircase for climbing on top of the tower. The portion between the inner and outer section in that case can be considered as solid one, defining the thickness of the tower walls. This thickness as per the mathematical relations established here, is half that of the diameter of the inner section.

Again, we may consider that the relationship expressed in the *sūtra* between the length (considering the meaning of *ayam* as length of the tower, as translated by Kangle) and the circumference (which bears a constant relation with the diameter), might have something to do with the stability of the structure.

It is well known that for the stability of any structure resting on a solid ground, the important parameters necessary to be considered are moment, vertical force, sectional modulus and horizontal force acting on the structure (these are well-known terms in civil and structural engineering, having mathematical connotations). Taking F as the horizontal force, h as height of the structure, D as the diameter of the structure, t as the thickness of the wall, P as the total vertical load at the base of the stucture, and Z as the sectional modulus of the base of the structure, the different parameters are defined as:

Moment $(M) = F * h,$

$Z = \pi D^2 t/4$

Area at the base $(A) = \pi D t.$

Now, for the stability of the structure $M/Z \angle P/A$

or, $Fh\ (\pi D^2 t/4) \leq P/\pi Dt$

or, $4Fh/\pi D^2 t \leq P/\pi Dt$

or, $4h/D \leq P/F.$

This, in other words, indicates that stability of such a structure against normal forces acting on it can be assessed in terms of the proportional relationship of the height and diameter. This is the modern practice in engineering of ascertaining the proportionality with respect to the forces, ascertained or assumed. Though not conforming exactly with the proportionality thus deduced by modern sophisticated calculations, *sūtra* 2.3.18 indicates the importance of proportionality of these two parameters. This knowledge must have been acquired at that time by observations, as in other cases discussed in this write-up. What attracts the attention is that identification of the participatory role of the components of a structure and their mutual relationship in ensuring the stability of a structure is well endorsed by the modern principles of structural analysis.

While on the subject, it will be of interest if we examine the discoveries in other fields of sciences and the time of their incorporation in the basic principles of design of civil engineering structures. A short summary of the same is listed below (Sen Sarma, 1979, pp. 121-2):

(a) Between 4500 and 2000 BC, suitability of clay as construction material, that is the plastic properties and mouldability of clay, use of straw, reeds, etc., with clay for improving mechanical properties to prevent crumbling; principle of independent horizontal members spanning vertical supports; durability and availability in shape and size of rocks with little or no further efforts for use as building stone; these leading to brick moulding and their use in buildings, post and lintel construction with rock slabs, etc., were discovered and used by early and hyperborean man.

(b) Up to AD 500, principles of arching, cantiliver action and stability of brick arches are to be found in the construction of vaulting of halls and similar structures in Mohammedan countries.

(c) Between AD 500 and 1000, usefulness of rocks of different varieties as building and decorative materials and pozzolana as cementing material, manifested in Roman structures.

(d) To AD 1700 — Behaviour of materials under static load, mode of fracturing and failure of lintels and beams when loaded, by Vitruvius; application of the principles of statics in structural analysis — systems of pulleys, levers, forces in arch, forces of triangulated structures by Leonardo da Vinci; behaviour of materials under tension, moment of resistance of beams proportional to their breadth, the square of their breadth and inversely proportional to their span; increase of the bending moment due to the weight of a beam proportional to the square of its length; properties of cylinders and hollow beams, by Galileo Galilei; laws of motion — the basis of the whole science of dynamics by Isaac Newton; laws of elastic properties of materials — *ut tensio sic vis* (as the pull, so the stretch), by Robert Hooke; all these discoveries influenced the planning and design of structures on mathematical basis with refinement, incorporating the results of testing materials by controlled experiments, which were evolved following these principles.

(e) To AD 1900 — Bending of beams of constant and variable sections under different conditions of loading, straight or curved initially, failure of short columns by compression and those more than an optimum length, by bending, by Daniel Bernoulli and Leonard Euler; laws governing the compression as well as tension under load, existence of shearing force, theory of earth pressure, stability of loose soils by C.A. Coulomb and W.J.M. Rankine; principles of structural theory, strength of materials, deflecting of beams under uniform and point loading, working theory of thin shells, arches, curved bars, etc., by C.L.M. Navier, Otto Mohr; problems of vibrations introducing new elements of dynamic forces to be taken into account for stability analysis by Lord Raleigh; cycloidal form of slip in soils — appoximate methods of analysis of stability of soil formations by Alexandare Collin; strain energy — that strain energy of any component in a loaded system may be shown in terms of horizontal and vertical displacements and angular relation, by Clark Maxwell and Alberto Castigliono, etc., etc., manifested in the application in design of the relationships of load, the stiffness of the column and itss length — evolving the column formula widely used in civil engineering, culminated in the birth of soil mechanics, application of the theory of moments in stability calculations, stability of slopes of soil, mechanism of landslides, stability of retaining structures, aseismic design of structures, etc.

Section C: Moving platforms (of the form of lift in a modern building)

While constructing the royal chamber with various means of emergency exit, in the form of concealed staircases, underground passages, etc., it has also been stipulated that the upper chamber immediately above the underground floor (meant as a shelter during any danger from outside) should be provided with a mechanism which would ensure quick and immediate descent from the room above into the underground chamber. The mechanism described in *sūtra* 1..20.2, as *yantrabaddha talabapatam* immediately conveys the idea of a platform — so designed and provided with mechanism that it is possible to lower it down with the king on it to the chamber below. This mechanism resembles the lift provided in a modern house for moving up and down, and which is propelled by pulley system. A pulley is one of the simple mechanical powers, consisting of a grooved wheel mounted in a block, so that a cord or the like may pass over it; used for changing the direction of power, especially for raising weights by pulling downwards, and vice versa. For pulling the platform upwards, the rope round the wheel requires to be pulled down, and for lowering, the rope wound round the wheel requires to be released. This system of working is also described in *sūtra* 12.5.45 — *yantraviśleṣaṇam va viśleṣyadhastadvapayeta*, that is for lowering down (the platform) something (obviosuly the rope round the wheel) should be mechanically loosened. The descriptions clearly indicate perfect working of such a platform. It can, hence, be inferred that the principle of putting the point of pull exactly along the line passing through the centre of gravity of the platform was known and perfected, probably through repeated trials and errors.

Section D: Spiral staircase

Sūtra 2.3.22 describes a staircase to be built for going up to the next higher floor, which according to the description *samucchrayad ardhatale stunabandhas ca* (2.3.21), means a mezzanine floor, built on pillars. Two alternatives have been indicated. The staircase has either to be concealed into the wall (*gūḍhabhitti sopānam*), or the one along which the ascent will be by circular movement from left to right (*bamataḥ pradakṣiṇa sopānam* — 2.3.22). The alternative suggestion arises from the use of the word *itarataḥ*, which also means 'otherwise than', 'different from', and also 'from or to another direction' (M.M. Williams-dictionary). Kangle in his translation has used the latter meaning (Kangle, pt. II, p. 66). This description is interesting. It is suggestive of the fact that the theoretical concept that vertical movement, or in other words, the gain of height from one level to another, achievable without any horizontal but by circular movement could be translated in practical construction. The description compares with sinistrally spiralled staircase, occupying much less horizontal space. The concept of saving the space, or the best utilisation of the same within confines of a room space is indicated when alternative of having a staircase embedded within the wall is given as a choice in place of the spiral one. The former, that is the concealed staircase has often been indicated in the same translation and comments, as a precautionary measure against enemies, but the suggestion of two types of staircases within the room, can very well be taken as an aesthetic concept of better utilisation of space for construction of service facility. A best room (*uttama-gṛha*) should be the one which is not clustered with or trammelled by such constructions and protrubations as would otherwise occupy a sizeable space of the room and would present unpleasant view. A spiralled staircase, occupying minimum horizontal space, or a staircase outside the floor space, concealed within the wall, will not hinder or interfere with the slickness and beauty of the room. Considering that *Arthaśāstra* displays deep sense of beauty and aestheticism in diverse spheres of life and living, this discourse may not appear to be baseless interpolation.

Appendix 2

Subject-wise Classification of Sūtras

[*Note*: Numbers before *sūtra*s indicate Book, chapter and *sūtra* number; e.g., 1.12.30 means Book 1, chapter 12 and *sūtra* no. 30 — as in pt. 2 of Kangle.]

A. Gems, minerals, ores and mining

ABOUT PEARLS AND CORALS

2.11.2: That from the Tāmraparṇi, that from Pāṇḍyakavata, that from the Pāśika, that from the Kula, that from the Cūrṇi, that from (Mt) Mahendra, that from the Kardama, that from the Srotasī, that from the Lake, and that from the Himavat, these are pearls.

2.11.3: The shell, the conch and miscellaneous are the sources (of pearls).

2.11.4: (A pearl that is) lentil-shaped, triangular, tortoise-shaped, semi-shperical, with a layer, coupled, cut-up, rough, spotted, gourd-shaped, dark, blue, and badly perforated, is defective.

2.11.5: (That which is) big, round, without a flat surface, lustrous, white, heavy, smooth and perforated at the proper place is excellent.

2.11.42: The coral from Alakanandā and from Vivarṇa, red and of

---

Note: *Sūtras* 2.11.29 to 2.11.32 and 2.11.34, 2.11.35, 2.11.36, 2.11.39, 2.11.42 demonstrate identification of gems by characteristic colours: *Sūtras* 2.11.2, 2.11.28, 2.11.37, 2.11.38, 2.12.17 indicate the geographical locations, sources of pearls and gems.

the colour of the lotus (is excellent), with the exception of that which is eaten by insects and which is bulging in the middle.

ABOUT GEMS

2.12.17: That which is clear, smooth, lustrous, possessed of sound, cold, hard and of a light colour is gem-ore.

2.11.28: Gems come from Koṭi, from the *mala*, and from beyond the sea.

2.11.29: The ruby, of the colour of red lotus, of the colour of saffron, of the (colour of) *pārijāta*-flower, (and) of the (colour of the) morning sun.

2.11.30: The beryl, of the colour of the blue lotus, of the *sirīsa*-flower, of the colour of water, of the colour of (green) bamboo, of the colour of the parrot's wing, yellow coloured, of the colour of cow's urine (and) of cow's fat.

2.11.31: The sapphire, with blue lines, like the *kalaya*-flower, deep blue, having the lustre of the *jambū*-fruit, having the lustre of the dark cloud, the 'delighter' (and) the 'streaming interior'.

2.11.32: The crystal, of the colour of *mulata*-flower, shedding a cool shower and the sun-stone.

2.11.33: Hexagonal, square or round, of a flashing colour, having a suitable form, clear, smooth, heavy, lustrous, with lustre inside and imparting lustre — these are excellences of gems.

2.11.34: With a dull colour and lustre, with grains, with a hole in the bloom, broken, badly bored, (and) covered with scratches, — these are blemishes.

2.11.35: The *vimalaka*, the *śasyaka*, the *añjanamūlaka*, the *pittaka*, the *śūlabhāka*, the *lohitākṣa*, the *mṛgasmaka*, the *jyotirasaka*, the

---

Note: *Sūtras* 2.11.33 and 2.11.34 indicate the general characteristics of excellent and inferior varieties of gems. Such characteristics for other individual gems are contained in *sūtras* 2.11.40 and 2.11.41 for diamond; 2.11.42 for coral.

maleyaka, the *ahicchatraka*, the *kūrpa*, the *pratikūrpa*, the *sugandhīkūrpa*, the *kṣīravāka*, the *sūkticūrṇaka*, the *śilāpravālaka*, the *pulaka* the *suklapulaka* — these are subsidiary types of (gems).

2.11.36: The rest are glass-crystals.

2.11.37: Diamonds come from Sabhārāṣṭra, from Tajjamarāṣṭra, from Kastīrarāṣṭra, from (Mt) Śrīkaṅṭaka, from (Mt) Maṇimanta, and from Indravana.

2.11.38: The mine, the stream, and miscellaneous are the sources.

2.11.39: Like Cat's eye, the *sirīsa*-flower, cow's urine, cow's fat, pure crystal, (or) the *mulaṭa*-flower, and of the colour of any one of the gems — these are the colours of diamonds.

2.11.40: (A diamond that is) big, heavy, capable of bearing blows, with symmetrical points, (capable of) scratching a vessel, revolving like a spindle and brilliantly shining is excellent.

2.11.41: That with points lost, without edges and defective on one side is bad.

ON METALS AND ORES

2.12.2: (Those liquids) that flow inside a hole, a cave, a table land at the foot, a rock-cut cave, or a secret dug-out in mountains whose regions are known (to contain gold-ore), that are of the colour of *jambū* — the mango — or the plam fruit or of a cross-section of ripe turmeric or of jaggery or of orpiment or red arsenic or honey or vermillion or white lotus or the feather of a parrot or a peacock, that have in their environs water and plants of the same colour, and that are viscous, clear and heavy, are gold-bearing liquids.

2.12.3: (If, when) thrown in water, they spread on the surface like oil and absorb mud and dirt, they are capable of transmuting copper and silver up to one hundred (times their own weight).

2.12.4: Exactly similar to that in appearance, (but) with a strong smell and taste, he should know to be bitumen.

2.12.5: Ores in the earth or rocks, which are yellow or copper coloured or reddish-yellow, which, when broken, show blue lines or are of the colour of the *mudga* or *māsa* bean or *kṣāra*, which are variegated with spots or lupms as of curds, which are of the colour of turmeric or myrobalan or lotus leaf or moss or liver or spleen or saffron, which when broken, show lines, spots or *svastikas* of fine sand, which are possessed of pebbles and are lustrous, which, when heated, do not break and yield plenty of foam and smoke, are gold ores, to be used for insertion, as transmuters of copper and silver.

2.12.6: (Ores) of the colour of a conch-shell or camphor or crystal or butter or a dove or a pigeon, or *vimalaka* (gem) or the peacock's neck, (or) of the colour of *saśyaka* (gem) or *gomedaka* (gem) or jaggery or unrefined sugar, (or) of the colour of the flower or *kovidara*, or lotus or *pātali*-flower or *kalaya*-flower or the flower of flax or of linseed, (and) containing lead, antimony, smelling like raw flesh, when broken (either) black with a white shimmer (or) white with a black shimmer, or all variegated with lines and spots, soft, (which) when being smelted do not split and yield plenty of foam and smoke, are silver ores.

2.12.7: In the case of all ores, when there is increase in heaviness there is increase in metal content.

2.13.3: That from the Jambū river, that from (Mt) Śatakumbha, that from Hāṭaka, that from Veṇu, that produced in Śṛṅgasūkti, that found in natural condition, that transmuted by means of liquids and that produced from mines — these are (types of) gold.

2.13.4: (Gold) of the colour of the lotus-filament, soft, smooth, not producing a sound and lustrous is best, reddish-yellow is of a middling quality, and red of the lowest quality.

2.13.5: Of the best (varieties), the pale-yellow and the white are impure.

2.13.10: That originating in Tuttha, that from Gauḍa, that from *kambu* and that from Cakravata, — these are (types of) silver.

2.13.11: (Silver which is) white, smooth and soft is best.

2.13.12: The opposite kind and that which tends to burst is bad.

2.12.12: Ore from rocks or a region of the earth, which is heavy, unctuous and soft (and which is) tawny, green reddish or red (in colour) is copper-ore.

2.12.13: That which is crow-black or of the colour of the dove or yellow pigment or studded with white lines (and) smelling like flesh, is lead-ore.

2.12.14: That which is grey like saline earth or of the colour of a baked lump of earth is tin-ore.

2.12.15: That which is made up mostly of smooth stones, is whitish-red or of the colour of *sinduvara*-flower is iron-ore.

2.12.16: That which is of the colour of *kakanda* ('crow's egg') or birch-leaf is *vaikṛntaka*-ore.

ON PURIFICATION OF ORES

2.12.8. Of these, those that are impure or dim in the interior flow in a pure form when infused in strong urine and caustic, when formed into lumps with (a paste of) the *rājavṛkṣa*, the banyan tree, the *pilu*, cow's bile, pigment and the urine and dung of the buffalo, the donkey and the young camel, either mixed with this (during boiling) or smeared with this (paste).

2.12.9: The insertion (in this ore) or the bulbous roots of the *kadalī* and the *vajra*, along with the caustic (of the ashes) of barley, *māsa*-bean, *sesamum*, *palāśa* and *pīlu* or along with the milk of cow and the goat, is productive of softness.

2.12.10: Honey and liquorice, goat's milk with sesamum-oil, mixed with ghee, jaggery and fermenting stuff, together with the *kandalī*, — with only three infusions in this, a (metal ore) that may have been broken even a hundred-thousand-fold becomes soft.

2.12.11: The insertion of (the powder of) the teeth and horns of the cow is the remover of softness (in metals).

2.13.6: He should cause that because of which it is impure to be removed by means of lead four times that quantity (*note*: to be read with *sūtra* 2.13.5 in connection with impure gold).

2.13.7: If it becomes brittle by the admixture of lead, he should cause it to be smelted with dried lumps of cow-dung.

2.13.8: If it is brittle because of (its own) roughness, he should cause it to be infused in sesamum-oil and cow-dung.

2.13.9: (Gold) produced from the mines, becoming brittle by the admixture of lead, he should turn into leaves by heating and cause them to be pounded on wooden anvils, or should cause it to be infused in the pulp of the bulbous roots of the *kadalī* and *vajra* (plants).

2.13.13: [*Note*: To be read with *sūtra*s 2.13.11 and 2.13.12 — about silver.] He should purify that with one-fourth part of lead.

2.13.14: That in which a crest has appeared at the top, which is clear, lustrous and of the colour of curds is pure.

METALLURGY

2.15.60: He should cause charcoal and husks to be taken to metal workshops and for plastering walls.

2.12.18: What is produced from ores, he (Director of Mines — *sūtra* 2.12.1) should put to use in factories for the respective metals.

2.12.23: The Director of Metals should establish factories for copper, lead, tin, *vaikṛntaka*, brass, steel, bronze, bell-metal and iron, also (establish) trade in metal ware.

2.12.24: The mint master should cause to be minted silver coins with one-fourth part copper (and) containing a hardening alloy of one *māsa* (in weight) of one of the following (viz.) iron, tin, lead and antimony, (of the denominations) of one *paṇa*, a half *paṇa*, a quarter *paṇa* and one-eighth *paṇa*; (further), copper coins with one quarter sustenance (of an alloy), (of the denominations of) one *māsaka*, a half *māsaka*, a *kakanī* and a half *kakanī*.

2.13.15: One *suvarṇa* of pure, turmeric-coloured (gold) is the standard.

2.13.16: Thence by substitution of one *kakanī* copper onwards (in a *suvarṇa* of gold) up to the limit of four (*māsaka*s), sixteen standards (are obtained).

2.13.44: In the work of the gold-plater, he should plate an article of copper with an equal amount of gold.

2.13.45: An article of silver, whether solid or hollow, he should cover with half the amount of gold.

2.13.46: Or, he should gild (the silver article) with one-fourth part of gold by means of the liquid or powder of sand-vermillion.

2.13.47: Ornamental gold of the best kind, possessed of excellent colour, passed through an equal amount of lead, turned into leaves by heating (and) made bright with Indus-earth, becomes the base of blue, yellow, white, green and parrot-feather colours.

2.13.48: And iron of the colour of the peacock's neck, white at the cut, shimmering (and) powdered after being heated, is for this (gold) a colouring matter one *kakanī* in measure of one *suvarṇa* (of gold).

2.13.49: Or, silver, purified, four times in a crucible of bones, four times in one with an equal amount of lead, four times in a dry crucible, three times in a pot-sherd (and) twice in cow-dung, thus passed through seventeen crucibles, (and) made bright with Indus-earth (is a base for colours).

2.13.50: From this one *kakanī* onwards up to two *māsa*s should be inserted in one *suvarṇa*; afterwards the adding of colouring matter (should follow); it becomes white silver.

2.13.51: Three parts of ornamental gold, strengthened with thirty-two parts of white silver, it becomes white-red.

2.13.52: It makes copper yellow.

2.13.53: Making the ornamental gold bright, he should give one-third part colouring; it becomes yellowish red.

2.13.54: (With) two parts of white silver, one part of ornamental gold produces the colour of the *mudga*-bean.

2.13.55: When smeared with half a part of black-iron, it becomes black.

2.13.56: Ornamental gold twice smeared with an enveloping liquid gets the colour of parrot's feather.

INDENTIFICATION AND TESTING OF GOLD: TESTING DEVICES

2.13.17: After first rubbing the gold (to be tested) on the touchstone, he should afterwards rub the standard gold (on it).

2.13.18: That with a streak of the same colour (as the standard) on places (on the stone) that are neither depressed nor elevated, he should know as properly tested; what is over-rubbed or lightly rubbed or powdered over with red chalk from underneath the finger-nail, he should know as deception.

2.13.19: Gold, touched with the forepart of the hand smeared with jati-varmillion or sulphate of iron dissolved in cow's urine, becomes white.

2.13.20: The streak (of gold) on the touchstone, that has filaments, is smooth, soft and lustrous is best.

TOUCHSTONES

2.13.21: Stone from the Kaliṅga or from the Tāpī, which is of the colour of the *mudga*-bean, is the best touchstone.

2.13.22: That which shows the exact colour is advantageous for either sale or purchase.

2.13.23: That which has the colour of an elephant, with a green tinge, is over-sensitive to colour (and) is advantageous for sale.

Note: *Sūtras* 2.13.44 to 2.13.56 relate primarily with plating, colouring of metals — and strictly do not fall under the category of metallurgy.

2.13.24: That which is firm, hard and of uneven colour, is not sensitive to colour (and) is advatageous for purchase.

MINING

2.12.1: The Director of Mines, being conversant with the science of (metal) veins in the earth and metallurgy, the art of smelting and the art of colouring gems, or having the assistance of experts in these, and fully equipped with workmen skilled in the work and with implements, should inspect an old mine by the marks of dross, crucibles, coal and ashes, or a new mine, where there are ores in the earth, in rocks or in liquid form, with excessive colour and heaviness and with a strong smell and taste.

2.12.20: He should make a miner who robs pay eight-fold, except in the case of jewels.

4.9.2: For an (officer) stealing an article of high value or gem from mines or factories for articles of high value, simple death (without torture, shall be the punshiment).

2.12.21: The thief and the person who lives (by mining) without permission, he should bind and force to work (in the mine), also the person who offers personal labour in place of fine. (compare *sūtra* 3.13.18 — one on whom a fine is imposed may pay off the fine by work).

2.12.22: He should let for part-share or on lease a mine that is burdensome in point of expenses or working; a light one, he should work himself.

4.1.51: In cases of information about mines, gems and buried treasure, the informants shall receive one-sixth part, one-twelfth part (if he be) a servant (of the state).

4.1.52: A treasure-trove over one hundred thousand (*paṇa*s) shall go to the king.

4.1.53: If less, he shall give one-sixth part (to the finder).

7.12.13: Of two mines, he who causes a mine to be opened, yielding plenty of valuable ores, with roads not inaccessible, and operated with small expenditure, over-reaches.

7.12.14-16: Even in that case, as between few things of high value and plenty of things of small value, 'The few things of high value are to be preferred, for, diamonds, rubies, pearls, corals and gold and silver ores swallow up the many objects of small value by their excessively high value', say the teachers. 'No', says Kauṭilya. A purchaser for an object of high value is rare and found after long time, while there are plenty for objects of small value because of a constant demand.

2.12.35-6: Thus, from the mines he should collect the price, the share, the surcharge, the monopoly tax, penalty, duty, compensation, fine, inspection fee, and manufacturing charges as well as the twelve kinds of metals and commodities (made from them). In this way he should fix the collection (of income) under various heads in the case of all commodities.

7.14.25: Mines are the source of implements of war; .....

1.12.37: The treasury has its source in the mines; from the treasury the army comes into being, with the treasury as its ornament.

2.12.29: Imported salt shall pay one-sixth part (as duty).

2.12.30: (Its) sale (shall be allowed) only after the share and the dues are paid (viz.), five per cent surcharge, the inspection fee and the manufacturing fee.

2.12.31: The purchaser shall pay the duty and a protective duty corresponding to the loss sustained by the king's goods; one who purchases at another place (shall pay) a fine of six hundred *paṇa*s in addition.

2.12.32: (A person selling) adulterated salt shall pay the highest fine (for violence), also the person living (by salt-manufacture)

Note: In consonance with the dictum in 7.12.16 — We find that considerable emphasis is given on common salt: *vide sūtra* 2.15.15 : Salt from the Indus-land, sea-salt, bida-salt, saltpetre, borax and salt from saline soil constitute the group of salts.

without permission, except forest-hermits.

2.12.33: Brāhmaṇas learned in Vedas, asceties and labourers may take salt for their food (without payment).

2.12.34: All other varieties of salt and sugar shall pay (only) the duty.

B. Land, agriculture, horticulture and irrigation

[*Note*: The patterns of settlement of lands are dealt with in details.]

2.1.7: He should grant (lands) to priest, preceptors, chaplain and Brahmins learned in Vedas (as) gifts to Brahmanis, exempt from fines and taxes, with inheritance, passing on the corresponding heirs, (and) to heads of departments, accountants and others, and to *gopas*, *sthānikas*, elephant-trainers, physician, horse-trainers and couriers, (lands) without the right of sale or mortgage.

2.1.8: He should allot to taxpayers arable fields for life.

2.1.9: Unarable fields should not be taken away from those who are making them arable.

2.1.10: He should take away (fields) from those who do not till them and give them to others.

2.1.11: Or, village servants and traders should till them.

2.1.12: Or, those who do not till should make good the loss (to the treasury).

2.1.13: And he should favour them with grains, cattle and money.

2.1.14: These they should payback afterwards at their convenience.

2.1.15: And he should grant to them favours and exemptions which would cause an increase in the treasury, (but) avoid such as would cause loss to the treasury.

2.1.17: He should grant exemptions at the time of settlement or as people come.

2.1.18: He should, like a father, show favours to those whose exemptions have ceased.

2.2.1: On land unsuitable for agriculture, he should allot pastures for cattle.

2.1.37: He should protect agriculture that is oppressed by the troubles of fines, labourers and taxes, and herds of cattle (oppressed) by thieves, wild animals, poison and crocodiles as well as by diseases.

2.1.20: He should cause irrigation works to be built with natural water sources or with water to be brought in from elsewhere.

2.1.21: Or, to others who are building (these), he should render aid with land, road, trees and implements, and (also render aid) to (the building of) holy places and parks.

2.1.22: If one walks out of the joint building of an irrigation work, his labourers and bullocks should (be made to) do (his share of) the work.

2.1.23: And he should be a sharer in the expenses and yet should receive no portion (of the benefits derived).

2.1.24: The ownership of the fish, ducks, and green vegetables in the irrigation works should go to the king.

2.2.2-6: He shall also allot land in forests to ascetics for Soma plantations and for Vedic learning. These shall be at least one *goruta* in extent and the safety of all movable and immovable things therein shall be assured.

He shall set-up a forest of similar size as a recreational forest for the king. Near the frontier or in any other suitable area, he shall set-up animal sanctuary where all animals are (welcomed as) guests.

He shall further demarcate forest areas, one for each kind of forest produce and set-up factories for goods made from such produce and create settlements of forests near these productive forests.

On the border of the kingdom he shall establish a forest for elephants, protected by other forests and under the supervision of the Chief Elephant Forester.

(*Sūtras* 2.2.2 to 2.2.6 — tr. by Rangarajan, pp. 181-2.)

2.24.1: The Director of Agriculture, himself conversant with the practice of agriculture, water-divining and the science of rearing plants, or assisted by experts in these, should collect, in the proper seasons, seeds of all kinds of grains, flowers, fruits, vegetables, bulbous roots, roots, creeper fruits, flax and cotton.

2.24.2: He should cause them to be sown in land, suitable for each, which has been ploughed many times, through serfs, labourers and persons paying off their fines by personal labour.

2.24.5: Sixteen *droṇa*s is the amount of rain in dry lands, one and a half times (that) in wet lands, where sowings are in conformity with (the nature of) the region, thirteen *droṇa*s and a half in the *asmaka*s, twenty-three in the *avantī*s, unlimited in the *aparānta*s and snowy regions, and (unlimited) as to time, in lands where sowings are made with the help of canals.

2.24.6: One-third of the (annual) rainfall in the first and the last months (together), two-thirds in the intervening two months, — this is the form of excellence (of the season).

2.24.7: Its ascertainment (is made) from the position, motion and impregnation of Jupiter, from the rise, setting and the movements of Venus and from modification in the natural appearance of the Sun.

2.24.8: From the Sun (is known) the successful sprouting of seeds, from Jupiter the formation of stalks in the crops, from Venus rain.

2.24.9: Three clouds raining (continuously) for seven days (each), eighty (clouds) showing drops of rain (and) sixty clouds accompanied by sunshine, this rainfall is even and beneficial.

2.24.10: Where it rains distributing wind and sunshine properly and creating three (periods for the drying of) cow-dung cakes, there the growth of crops is certain.

2.24.11: In conformity with that, he should cause crops to be sown, requiring plenty of water or little water.

2.24.12: *Śāli*-rice, *vṛhī*-rice, *kodrava*, sesamum, *priyaṅgu*, *udaraka* and *varaka* are the first sowings.

2.24.13: *Mudga*, *māsa* and *saimbya* are the middle sowings.

2.24.14: Safflower, lentils, *kulattha*, barley, wheat, *kalaya*, linseed and mustard are the last sowings.

2.24.15: Or, the sowing of seeds (should be) in conformity with the season.

2.24.16: What is left over from sowing, farmers cultivating for half the produce should till, or those who live by personal labour (should work it) for a one-fourth or one-fifth share.

2.24.17: The (farmers) shall pay a share for uncultivated land as desired (by the king), except in distress.

2.24.18: The (farmers) shall pay a water rate of one-fifth in the case of water set in motion by the hand from their own water-works, one-fourth when set in motion by shoulders, and one-third when set flowing in channels by mechanism, one-fourth when lifted from rivers, lakes, tanks and wells.

2.24.19: According to the amount of water (available) for the work, he should decide on wet crops, winter crops or summer crops.

2.24.20: *Śāli* and others are the best (crops), vegetables midling, sugar-cane worst.

2.24.21: For, sugar-canes are fraught with many dangers and require (much) expenditure.

2.24.22: (A region) where the foam strikes (the banks) is (suited) for creeper fruits, (regions on) the outskirts of overflows, for long pepper, grapes, and sugar-canes, (those on) the borders of wells, for vegetables and roots, (those on) the borders of moist beds of lakes, for green grasses, ridges for plants reaped by cutting, (such as) perfume-plants, medicinal herbs, *usira*-grass, *hrībera*, *piṇḍaluka* and others.

2.24.23: And on lands suitable for each, he should raise plants that grow on dry lands and that grow in wet lands.

2.24.24: Soaking in dew (by night) and drying in the heat (by day) for seven days and nights (is the treatment) in the case of seeds of grains, for three days and nights or five in the case of seeds of pulses, smearing at the cut with honey, ghee and pig's fat, mixed with cow-dung in the case of stalks that serve as seeds, (smearing) with honey and ghee, in the case of bulbous roots, smearing with cow-dung in the case of stone-like seeds, (and) in the case of trees, burning in the pit and fulfilment of the longing with cow-bones and cow-dung at the proper time.

2.24.25: And when they have sprouted, he should feed them with fresh acrid fish along with the milk of *snuhi*-plant.

2.24.27: However, at the first sowing of all kinds of seeds, he should sow the first handful (after it is) immersed in water containing gold and should recite the following *mantra* : 'Salutation to Kaśyapa, the lord of creation and to the god (of rain) always. May the divine Sītā prosper in my seeds and my grains.'

3.9.27: In the case of damage to the ploughing or seeds in another's field by the use of a reservoir, channels or a field under water, they shall pay compensation in accordance with the damage.

3.9.28: In case of mutual damage to fields under water, parks and embankments, the fine (shall be) double the damage.

3.9.29: A tank on a lower level, constructed afterwards, shall not flood with water a field watered by a tank on higher level.

3.9.30: A (tank) constructed on a higher level shall not prevent the flooding with water of a lower tank, except when its use has ceased for three years.

3.9.31: For transgression of that, (the punishment shall be) the lowest fine for violence and emptying of the tank.

3.9.32: The ownership of a water-work, not in use for five years, shall be lost except in cases of distress.

3.9.33: When tanks and embankments are newly constructed, an exemption (from taxes) for five years (should be granted), when those that are ruined and abandoned are renovated, an exemption for four years, when those that are over-grown with weeds are cleared, for three years, when dry land is newly brought under cultivation, for two years.

3.9.34: He is free to mortgage or sell.

3.9.35: (Owners) may give (water) in return for a share of produce of various kinds from sowings in fields, parks and gardens by (their) dug-out channels, structures based on rivers or tanks, or to others as it may be advantageous.

3.9.36: And those who use these on lease, on hire, as a pledge, for a share or with authorisation for use, shall keep (them) in repair.

3.9.37: In case of failure to repair, the fine is double the loss.

3.9.38: For one letting out water from the dams out of turn, the fine shall be six *paṇas*, also for one obstructing, through negligence, the water of others when it is their turn.

3.10.1: If one obstructs a customary water-course in use or makes (a new one) that is not customary, the lowest fine for violence (shall be imposed), also if one constructs in another's land a dam, a well, a holy place, a sancturary or a temple.

3.10.2: If a (person) himself or through others puts to mortgage or sale a charitable water-work, continued since old times, the middle fine for violence (is to be imposed), the highest on witnesses, except when it is in ruins and abandoned.

3.10.8: If the owner of the field takes away the field or the tenant leaves it at the time of sowing, the fine shall be twelve *paṇas*, except in cases of calamity or unbearable conditions.

3.10.14: If one does not till land that is inalienable, another may use it for five years and return it after receiving compensation for his exertions.

6.1.8: Describes the attributes which render excellence to a country: '. . . deviod of mud, stones, salty ground, uneven land, . . . endowed with agricultural land . . . with protected pastures, not depending on rain for water, ...'

7.12.4: Of two water-works, one with naturally flowing water is preferable to one into which water has to be brought.

7.12.5: Even among two water-works with naturally flowing water, one with an extensive region for sowing in preferable.

C. CIVIL ENGINEERING AND ARCHITECTURE

1.20.1: On a site recommended by experts in building, he should cause the royal residence to be built, with a rampart, a moat and gates and provided with many halls.

1.20.2: He should cause to be constructed a living chamber in the centre in accordance with procedure laid down for the treasury, or a maze-house with concealed passages in walls and in its centre a living chamber, or an underground room with its opening covered by the wooden image of a deity in a nearby sanctuary and having many subterranean passages (and) above it a palace with a staircase concealed in a wall or having an entrance and an exit through a hollow pillar as a living chamber with the floor fixed to a mechanism (and thus) capable of sinking below, in order to counteract a calamity or when a calamity is apprehended.

1.20.10: In a part of the apartments at the back (should be) ladies' rooms, establishments for maternity and sickness and places for trees and water.

1.20.11: Outside (that) should be quarters for princesses and princes.

1.20.12: In front (should be) the dressing room, the council chamber, the assembly hall and a place for the ministers in charge of princes.

1.20.13: In the (open) spaces between the apartments, the palace guards should be stationed.

2.3.1: In all four quarters, on the frontiers of the country, he should cause a nature-made fortress, equipped for fight, to be made; a water fort (either) an island in the midst of water or high land shut in by water, (or) a mountain for (either) consisting of rocks or a cave, (or) a desert fort (either) one without water and shrubs or a salty region, or a jungle fort (either) a marshy tract with water or a thicket of shrubs.

2.3.3: In the centre of the country, he should lay out a *sthānīya*, the headquarters for revenue, on a site recommended by experts in the science of buildings, at the confluence of rivers or on the bank of a lake that never dries up, either a (natural) lake or a (man-made) tank, round, rectangular or square or in accordance with the nature of the building site, with water flowing from left to right, a market town, served by a land-route and a water-route.

2.3.4: He should cause three moats to be dug round it, at a distance of one *daṇḍa* from each other, fourteen, twelve and ten *daṇḍa*s broad, three-quarters or half of the breadth deep, one-third (of the surface breadth) at the bottom or square with the bottom, paved with stones or with the sides (only) built of stones or bricks, reaching down to (natural springs of) water or filled with water coming from elsewhere, with (arrangements for) draining excess water, and stocked with lotuses and crocodiles.

2.3.5: At a distance of four *daṇḍa*s from the moat, he should cause a rampart to be made out of the earth duge out, six *daṇḍa*s high, made compact, twice that is breadth, piled upwards with a platform-like (flat) surface (at the top) or with a jug-like side, pounded by elephants and bullocks (and) having (on the sides) clusters of thorny bushes and poisonous creepers.

2.3.6: With the remainder of the earth, he should cause the low places in the residential areas or palace grounds to be filled up.

2.3.7: On the top of the rampart, he should cause a parapet to be built, double the breadth in height, built of bricks, from twelve *hasta*s upwards up to twenty-four *hasta*s, either odd or even in numner (of *hasta*s in height), with a passage for the movement of chariots, shaped like a palm-stem and with top decked with 'drums' and 'monkey-heads'.

2.3.8: Or, he should cause it to be made of stones, close-knit with big slabs, but under no circumstances (should he have it) made of wood.

2.3.9: For, fire remains lurking in it.

2.3.10: He should cause turrets to be made, square with the breadth, provided with steps for going down, (of the) same (length) as the height, and at a distance of thirty *daṇḍa*s from each other.

2.3.11: Midway between (every) two turrets, he should cause a tower to be built, with two stories inclusive of a hall, (and) one and a half times in length.

2.3.12: Between each turret and tower, in the centre, he should cause to be erected a board, compact with planks having holes with coverings, as a place (from which to fight) for three archers.

2.3.13: In the intervening spaces, he should cause a 'gods' way' to be made, two *hastas* in breadth and four times that in length, at the side.

2.3.14: He should cause tracks to be made of a width of one *daṇḍa* or two *daṇḍa*s, and in an unassailable place a runway and exit-door.

2.3.15: Outside (the fort), he should cause a covered road to be made that is strewn with knee-breakers, a mass of pikes, pits, concealed traps, barbed wires, 'serpent-backs', 'palm leaves', 'three peaks', 'dog's jaws', bars, 'jumping sandals', frying pans and ponds.

2.3.16: Having made on both sides of the parapet a 'ram's head' one *daṇḍa* and a half (in extent), he should cause the gateway to be laid out, with space for six beams of the tower, from five *daṇḍa*s (square) onwards increased by one *daṇḍa* up to eight *daṇḍa*s square, or one-sixth part more than the length or one-eighth part (more).

2.3.17: From fifteen *hasta*s onwards increased by one *hasta* up to eighteen *hasta*s (should be) the height of the floor.

2.3.18: The circumference of a pillar (should be) one-sixth of the length, double that the portion fixed in the ground; the capital (should be) one-fourth (the length) in circumference.

2.3.19: Occupying one-fifth part of the ground floor (should be) a hall, a well and a border room.

2.3.20: There should be two platforms facing each other, occupying one-tenth part (of the space) and in between two doors and a room.

2.3.21: And at half the height of the floor, there should be a structure with pillars.

2.3.22: There should be an upper chamber measuring half the built-in area or having a space one-third of it, with sides constructed with bricks, with a staircase on the left running from left to right and on the other side a staircase concealed in the wall.

2.3.23: The crest of the arch should be two *hasta*s (in height).

2.3.24: The two door-leaves should occupy three-fifths (of the space).

2.3.30: The bridge should be equal (in width) to the opening and capable of being withdrawn, or it should be made of earth where there is no water.

2.3.31: Having fixed an opening equal to (the width of) the parapet, he should cause the *gopura* (gate) to be made, with one-third shaped like a lizard's mouth.

2.3.32: Having made a well in the middle of the parapet, he should erect the *puṣkariṇī*-gate; with four halls at a distance of one (*daṇḍa*) and a half (from each other) and with holes, the *kumārīpura*-(gate); a bare house with two stories, as the *muṇḍaka*-gate; or (he should erect gates) in accordance with the (availability of) land and materials.

2.4.1: Three royal highways running west to east and three running south to north, that should be the division of the residential area.

2.4.2: It should have twelve gates (and be provided) with suitable (places for) water, drains and underground passages.

2.4.3: Roads should be of a width of four *daṇḍa*s.

2.4.4: The royal highway and roads in *droṇamukha*, a *sthānīya*, the countryside and pasture lands as well as paths in a harbour town, a battle array, a cremation ground and a village (should be) eight *daṇḍa*s (in width).

2.4.5: Paths on irrigation works and in forests four *daṇḍa*s, paths for elephants and along fields two *daṇḍa*s, five *aratni*s the chariot path, four the cattle-path (and) two the path for small animals and men.

2.5.1: The Director of Stores should cause to be built a treasury, a ware-house, a magazine, a store for forest produce, an armoury and a prison-house.

2.5.2: Having caused to be a four-cornered well free from water and dampness, and having paved the walls on both sides and the bottom with big slabs of stone, he should cause an underground cellar to be made with a framework of strong timber, level with the ground, with three floors, with various arrangements, with ground and upper floors having well-made surfaces, with a single door, (and) with a staircase provided with a mechanism.

2.5.3: On that he should cause a treasure-house to be made with a prohibition (to enter) on two sides, with an entrance-hall, built of bricks, and surrounded by (channels) storing goods;

or (he should cause to be built only) an (above the ground) palace.

2.5.5: He should cause to be made a warehouse and a magazine with pillars of baked bricks, with four halls (opening on to a quadrangle), with many, ground and upper floors, (and) with a (secret) passage for escape through hollow pillars on two sides, in between, a store for forest produce with many long halls (and) with the walls surrounded by rooms, a similar (structure) with an underground cellar as an armoury, (and) separate prison-houses for those convicted by judges and for those punished by high officers, with separate apartments for women and men (and) with their rooms well-guarded against escape.

2.5.6: And in all (structures) he should cause the halls to be provided with a ditch, a well, a latrine and a bath-room, with precautions against fire and poison, with protection by means of cats and ichneumons and with (arrangements for) worship of their respective deities.

3.8.13: Between all two structures or two projecting rooms, (there is to be) an open lane one *kisku* (wide) or three *pada*s.

3.8.14: Between them, the distance between the eaves of roofs (is to be) four *aṅgula*s, or one may over-lay the other.

3.8.15: He should cause to be made a side-door in the intervening lane, measuring one *kisku*, for making repairs to what is damaged, not (allowing) crowding.

3.8.16: For light, he should cause a small window to be made high up.

3.8.19: And he should cause that part above the verandah which requires protection, to be covered by matting, or a wall touching (the roof), for fear of damage by rain.

D. Environment, wild and pet animals, flora, civic sense

PROTECTION TO ANIMALS: WILD AND DOMESTICATED

2.26.1: The Supervisor of slaughter should impose the highest fine (for violence) for binding, killing or injuring deer, beasts, birds or fish for whom safety has been proclaimed and who are kept in reserved parks, the middle fine on householders (for these offences) in reserved park enclosures.

2.26.2: For binding, killing or injuring fish and birds whose slaughter is not current, he should impose a fine of twenty-six *paṇa*s and three quarters, (for binding) deer and beasts, double (that).

2.26.5: Sea fish having the form of an elephant or a horse or a man or a bull or a donkey, or those from lakes, rivers, tanks or canals, curlew, osprey, gallinule, swan, ruddy goose, pheasant, *bhṛṅgarāja*, *cakora*, *mattakokila*, peacock, parrot and *madanasārikā*, which are birds for sport and auspicious (birds), also other creatures (whether) birds or deer, should be protected from all dangers of injury.

2.26.6: For transgression of (this) protection, the first fine for violence (shall be imposed).

2.26.10: The calf, the bull and the milch-cow among these (animals) are not to be killed.

2.26.11: For one killing (them, there shall be) a fine of fifty *paṇa*s, also for (one) torturing (them) to death.

2.26.14: Beasts, deer, wild animals and fish, belonging to sanctuaries, should, when harmful, be killed or bound to places other than the place of their protection.

2.29.16: He who himself kills (an animal) or incites another to kill or steals or incites another to steal shall be executed.

2.29.20: Cowherds shall (care for and) treat the young, the old and the diseased.

2.29.21: They shall graze (the cattle) in a forest from which the fear of danger from theives, wild animals and enemies has been removed by fowlers and hunters and which is sutiable in different seasons.

2.29.22: And in order to frighten serpents and wild animals and to know the movements in the pastures, they should tie a bell round the necks of timid animals.

2.29.23: They should take (cattle) down to water, to which the descent is even and broad, and which is free from mud and crocodiles, and should guard them (at the time).

2.2.4: And he should establish on its border or in conformity with the (suitablitiy of the) land, another animal park where all animals are (welcomed) as guests (and given full protection).

2.2.6: On the border (of the kingdom), he should establish a forest for elephants guarded by foresters.

2.2.7: The Superintendent of the elephant-forest should, with the help of guards of the elephant forest, protect the elephant forest (whether) on the mountain, along a river, along lakes or in marshy tracts, with its boundaries, entrances and exits (fully) known.

2.2.8: They should kill anyone slaying an elephant.

[Chapters 2.30 and 2.31 deal with the care to be taken, including stables, food to be given, etc., for horses and elephants — in great detail.]

3.19.26: For causing hurt to small animals with wood and other things, the fine shall be one *paṇa* or two *paṇa*s, double that for causing bleeding.

3.19.27: For these same offences concerning big animals, the fine shall be double and (payment of) expenses for treatment and cure.

FLORA

3.19.28: For cutting the shoots of trees in city parks that bear flowers or fruit or yield shade (the fine shall be) six *paṇa*s, for cutting small branches twelve *paṇas*, for cutting stout branches twenty-four *paṇas*, for destroying trunks the lowest fine for violence, for uprooting (the tree) the middle (fine).

3.19.29: In case of bushes and creepers bearing flowers or fruit or yielding shade the fines shall be half, also in case of trees in holy places, penance-groves and cremation grounds.

3.19.30: In case of trees at the boundaries, in sancturaries, and of trees that are prominent, these same fines doubled shall be imposed, also (in case of trees) in royal parks.

CIVIC SENSE, AWARENESS AND RESPONSIBILITIES

[These aspects have been summarised below. Violation of most of these conventions is punishable offence, details of punishments specified have been omitted. From the nature of violations attracting punishments, the obligations are indicated: References — Rangarajan, *op. cit.*, pp. 372-6.]

General responsibilities

- Causing injury by the collapse of a rickety dwelling or cart; an unsupported pillar or beam; an unsheathed weapon; an uncovered or concealed pit or well. — 4.13.13
- Not coming to the rescue of someone being hurt by his horned or tusked animal normally, or if failing to respond to cries for help. — 4.13.18

Traffic code

- Reckless driving of a cart except when it is due to an unforeseeable accident causing injury to human beings or animals, or causing death to animal. — 4.13. 23-4
- Owner of a cart or accompanying adult is to be punished or cart confiscated if there is no adult driver, or driver is a minor or in case of an unaccompanied minor in a cart. — 4.13.25-6

Fire prevention

- Lighting fire inside the house during the two middle quarters of the day. — 2.36.16
- Not providing fire fighting equipment. — 2.36.18
- Not hastening to save his own house on fire, if he is the owner of the house or if he is the tenant. — 2.36.23
- Letting a house catch fire through negligence — 2.36.24, or arson (deliberately setting fire). — 2.36.25.

Hygiene

- Throwing dirt on the road, blocking it with mud or water. — 2.36.26
- Throwing dirt on the royal highway, blocking it with mud or water. — 2.36.27
- For using a holy place, a water reservoir, a temple, a royal building, for using as urinal/latrine. — 2.36.28
- For throwing dead bodies of cat, dog, mongoose, snake, donkey, mule, camel or cattle, human crops inside the city. — 2.36.30
- For taking out a dead body by an unauthorised gate. — 2.36.31
- Burial/cremation in a place other than the designated grounds. — 2.36.33
- A gatekeeper who lets these happen. —2.36. 32

Damage to public property

- Damaging plants and trees in city parks, sansctuaries, holy places and cremation grounds, particularly those which bear fruits or flowers or provide shade, which include such offences as damaging a sprout, branches of any size, cutting off trunks, uprooting (including those which mark boundaries, are objects of worship or those in royal enclosures). — 3.19.28-30

Damage to private property

- Immovable property : damaging wall of someone else's house by hitting it, breaking or demolishing a wall. — 3.19.24
- Throwing damaging things or things which may endanger life, into another's house. — 3.19.25
- Breaking open the sealed door of a house. — 3.20.15

House building

- Without adequate rainwater drainage. — 3.8.23
- Causing damage to the wall of another house by letting water collect. — 3.8.22
- Causing damage to another house by letting urine or dung collect. — 3.8.22
- Having a cart shed, animal shed, fireplace, water jar, grinding mill or pounding mortar too near another house. — 3.8.12
- Having a fresh water course too near a neighbour's house. — 3.8.10
- Having a dung hill, a sewage channel or a well too near a neighbour's house as to cause nuisance. — 3.8.7
- Neighbour's verandah needing protection from rain not protected; house built with a door or window directly facing a door or window of a neighbour's house. — 3.8.20
- Causing obstruction and preventing the enjoyment of others. — 3.8.21

Tenancy

- Tenant who refuses to leave after being asked to or landlord who evicts, without due cause, a tenant who has paid his rent. — 3.8.24

Common facilities

- For not helping to build common facilities, for obstructing their use, or for preventing their rightful use. — 3.8.26
- For destroying a common facility. — 3.8.27
- Causing harm to entire neighbourhood. — 3.20.15
- Causing injury (by collapse of) unstable house. — 4.13.13

Privacy and mutual help

- Interfering in a neighbour's affairs without reason, not running to help a neighbour in distress. — 3.20.16

Appendix 3

Deconstruction of a few sūtras

ANY critical study of literary texts translated from one language to other poses several difficulties, especially when such translations generate questions about their adequacy and appropriateness in expressing reality. Every language has its particularity and applicability mostly related to social culture, behaviour, habits, rituals — in short, the social set-up and environment. Apart from this, any attempt to render into anothr language the real and appropriate version of any text dealing with specialised topics, demands adequate proficiency in those subjects. Without that, the literal translation becomes not only confusing but often conveys meanings verging to absurdity. Exponents like Kangle is very conscious about this aspect when he stated that

> The *Arthasastra* is admittedly a difficult work, and very often it is not possible to be quite sure of the exact meaning of the author.
>
> —Kangle in the Preface to the First Edition, 1963

In this context he states that

> It is in the vocabulary, however, that the language of this work shows striking peculiarities. There are numerous technical terms. Many of them are defined in the work itself, but there are others that are not defined. . . . It is true that we cannot be quite sure of the meaning of many words and expressions in this text. So, too the interpretation of many passages in the text must remain uncertain. . . .
>
> — Kangle, pt. III, pp. 38-9

Such interpretations, taken from out of multi-meaning of any word the most appropriate one with reference to the context of the subject dealt in any *adhikaraṇa* (Book) or a *prakaraṇa* (section of a book), may bring out the intended meaning. This must have been followed in the translations, but departures are also noticed. An example may be given of the translation of the *sūtra* 2.24.5 by Kangle (pt. II, p. 149), and by Pran Nath (*Indian Antiquary*, June, 1931, p. iii). The *sūtra* reads

> *ṣoḍaśadroṇam juṅgalanam vărṣăpramāṇam, adhyadharmānūpatam, ardhyatrayodaśamakanam, trayovinśatiravantinam, amitamparāntānām haimanyānām ca, kulyāvāpānam ca kālataḥ* ।

The *sūtra* occurs in the section under the heading of Director of Agriculture, and dealing with the practice of agriculture, in which the condition of the land, sowing time and practice, climatic conditions including water availability from rains, sunshine, etc., have been described. The *sūtra* has been translated by Kangle as

> sixteen *droṇa*s is the amount of rain in dry lands, one and a half times (that) in wet lands, where sowings are in conformity with (the nature of) the region, thirteen *droṇas* and a half in the Asmakas, twenty-three in the Avantis, unlimited in the Aparantas and the snowy regions, and (unlimited) as to time, in lands where sowings are made with the help of canals.

Pran Nath, on the other hand, considers *varṣă-pramāṇa* meaning 'the standard of the produce for the year', even though it 'may mean a rain-gauge'. 'The *droṇa*s in the passage', he considers as 'refer to the standard share of produce fixed for the crown lands situated in the different countries'. Further, according to him, the meaning of the following words appearing in the *sūtra* should be as — (i) *haimaṇyānām* — wintry or cold, and not Himālaya; (ii) *kulyavāpa* — crop irrigated by well, tank or water reservoir, (iii) *amitam aparāntanaṃ* — the quantity of produce of the *aparāntajanapada*s is not known (or not ascertained). He, accordingly translates the whole passage as:

> The annual measure (of produce to be taken as the king's

> due) is sixteen *dronas* in the country of Jangala; twenty-four *dronas* in moist (marshy or low) countries fit for agriculture; thirteen and half *dronas* in the *janapadas* of Asmaka; twenty-three *dronas* in Avanti; the quantity in Aparanta is not measured; the crop grown in winter and irrigated (by well, pond, tank, lake, etc.), should be ascertained according to the time.

Though not very certain about the correct interpretation, he seems to establish that

> Kautilya's king possessed landed property in the following countries: Aparanta, Asmaka, Avanti, Jangala and Anupadesa.

Needless to point out that this translation certainly strikes a discordant tone when the preceding and succeeding *sūtras* are scanned.

The example has been cited only to bring home the point that when considered in the context of the subject being dealt in any section, it is possible (and also advisable) to extract the real meaning, tacitly expressed by some words after carefully selecting the approved alternative meaning of those. This method has been adopted in re-interpreting some of the *sūtras*. In doing so, it has been considered that *Arthaśāstra* is mostly written in prose of *sūtra* form. *Sūtra* is not always grammatically complete with all the components of a sentence, but only a 'short or concise sentence used as memorial rule': *svalpākṣaramasaṁdigdham saravad viśvatomukham: aṣṭobhamanavadyam ca sūtram sūtravido viduḥ*. Accordingly *sūtras* offer scope of alternative interpretations; but when done with reference to the context, such interpretation is likely to bring out the most acceptable suggestive meaning.

1. ***Sūtra* 1.20.2**: The *sūtra* describes the various rooms to comprise the royal residence. These rooms are living room, underground chamber, a maze room (*mohanagṛha* — which according to Meyer is 'a love chamber', but this meaning has been ruled out by Kangle). In this translation, Kangle

has accepted them to be alternative to one another, by inserting the word 'or' between the description of each type of chamber. 'Or' obviously is used for the Sanskrit word *va*. The letter also means 'and', 'as', 'also' (Apte, p. 839), in which case each chamber assumes separate entities to be constructed as components of the royal residence, without indicating one as the alternative or substitute of the other.

Meyer's translation of *mohanagṛha* in the sense of 'love chamber' may not be discarded without ascribing the reason for it. Associating the concealed doors and stairs with under-ground chamber, and perhaps laying stress more on the question of safety and security of the king, this word has been translated as 'maze house' by Kangle. But Meyer's translation assumes reality and strong endorsement when we look into the prescribed arrangements of other rooms, appearing most sophisticated even in terms of modern architectural concepts, and the daily routine of the king, divided between the royal responsibilities and his private time. While the king has to adhere strictly to royal duties during specified hours of the day and night, 'there is no need for such a king to deprive himself of all sensual pleasures (and lead a life of austerity) so long as he does not infringe his *dharma* or harm his own material well-being' (1.7.3, Rangarajan, p. 145). King's daily routine has been prescribed in details in *sūtra*s 1.19.9 to 1.19.25., in which 'the king is allowed 10-1/2 hours out of 24 hours as his personal time — three hours for bath and meals, one and a half hours for recreation and six hours at night, during which he can sleep four and a half hours' (Rangarajan, p. 148). In consonance with other forms of utilitarian space within the royal residence, concept of a love chamber and all the other types of chambers does not seem to be out of tune with other aristocratic and sophisticated concepts of lifestyle manifested by the space allocations within a royal residence, like bedroom, dressing room, room for sick or maternity room, rooms allocated separately for prince and princess, and so on.

2. **Sūtra 1.20.4**: The last part of the *sūtra* — *vaidyutena bhasmana mṛtsanjuktena karakavārinavaliptam ca,* is interpretable in altogether different way, when the cloaked meanings of some of the words are extracted for understanding the nature and properties of some construction materials. The word *bhasmana* means, that which has been pulverised or calcined by fire, and *vaidyuta* carrying the meaning of intense fire (like that induced by lightning). The material that is calcined is clay (*mṛt*). The meaning of the word *karakavāri*; as applied here is not clear from the translation in which it is stated to be 'hail water'. Unless it means as quenching with ice-cold water, the meaning can be sought with other usage of the word *karaka* and *vāri*. The former means a water vessel. Nearest to this are words like *kadaka* which means 'sea-salt obtained by evaporation', and *vāri* also means a kind of salt. The two taken together may point out to such a salt, composed of calcium, potassium, and magnesium, which constitute the ingredients of cement. Transposing the words '*vaidyutenabhasmana mṛt karakavārina sañjuktenavaliptam*', the last part of the *sūtra* means 'plastering with the admixture of calcined clay (pumice brick) and salts obtained from sea water (by evaporation)'. The possibility of such a mixture indicating some form of pozzollana has been discussed in the text. *Vaidyuta bhasma* as preventor of fire has also been pointed out in *sūtra* 14.2.31.

3. **Sūtras 1.20.10 and 11**: Re-arrangement of these *sūtras* taken together would convey better meaning. *Sūtra* 10 has terminated with *vṛkṣadokasthānam ca*, and *sūtra* 11 reads as *vahiḥ kanyākumārapuram*. The two may be edited as *pṛṣṭataḥ kakṣavivāge śrīniveśa garbhabyādhi saṁsthā, vṛkṣadokaṣṭhānām ca vahiḥ kanyākumārapuram*. This may be translated as 'In the placement (layout) of rooms, the maternity room and sick room, should be at the back, and outside the quarters for prince and princess, there should be places for plantation (garden) and water (artificial water storage tanks/cisterns).' This conveys a more apposite and coherent meaning.

4. **Sūtra 2.1.20**: *sahodakamāhāryodakam vā setum vandhaet — setu* means embankement, raised ridge of earth. The *sūtra* means creation of reservoir, either around the existing water body (*sahodakam*), which may be a tank or some depression capable of holding water, made favourable to do so, or by diverting water from other perennial sources. The act indicates creating artificial reservoirs for harnessing rainwater or available water resource, preventing loss by run-off, or transferring water on to a region with favourable ground condition for holding water but not endowed wtih perennial source. The modern art of irrigation is based on this principle. Distinctions have been made between *sahodaka* (natural body of water) and *ahāryodaka* (water transferred from elsewhere), either by construction of dam or embankment (*sūtra* 7.12.4). The former is adjudged as a better choice than the latter. Distinction has also been made between two types of constructions as embankment (*setu*) and dam (*bandha*). That these two are separate types is indicated by the use of dual number between *setu* and *bandha*. Hence, these two words should be taken in these senses throughout the text. *Setu* has been used in the sense of embankment or an earthen ridge around depressions or around cultivated fields in *sūtra*s 2.35.3 and 3.9.32, and the respective uses and utilities of these types of constructions have been distinguished elsewhere also, in irrigational practice.

5. **Sūtra 2.3.5**: The *sūtra* describes the construction and form of an earthen rampart four *daṇḍa*s away from a moat. The essential part of it, the pillar-like high-rise mound is constructed, after having ensured that the foundation is made stable enough to support the above super outfit. In this, *avarudham* points to underground part of the structure, and the depth of this enclosure has been prescribed as *saddaṇḍa ucchṛtam*. Though *ucchṛtam* means raised (normally high above), this word used for indicating the enclosed foundation below ground, may in all fairness be taken of the depth with reference to the lowest level of the foundation.

The superstructure, the piled up mound, serving as supporting prop/pillar (*viṣkambha*) would rise above the ground twice as much as the depth, that is twelve *daṇḍa*s. This interpretation, using approved alternative meanings of several words, offers a clear picture of the concept and process of civil construction. Lack of proper understanding of the purported meaning of *sūtra*s in the context of the subject they deal with, leads to involved and often abstruse explication. Next sentence independently conveys clearer meaning, with, of course, introducing a silent verb in the sentence. This practice is in vogue, as elsewhere, in the treatise. The latter part should mean that the earthen pile should be compacted after the earth is pounded by elephants or bulls, should have a platform at the top, its mid-region should have outward bulge, (might be to make it difficult for scaling the wall), and thorny bushes and poisonous creepers should be planted on its surface.

6. ***Sūtras* 2.3.10-2.3.13**: The *sūtra*s describe various constructions at or associated with the rampart round the fort. These are all meant for providing protection to the fort inside the rampart, and are strategically located for the purpose. Words in the *sūtra*s used for describing the structures are found to have alternative meanings, and actual meaning of these are desirably to be selected in accordance with the overall description of the layout plan of the structures and in the context of their functional and utilitarian concepts. The discrimination of such meanings are necessary to avoid confusion. The approved alternative meanings of some of the words appearing in the *sūtra*s, adopted for their reinterpretation are as below:

   I. *viṣkambha*: pillar;

   II. *aṭṭālaka*: *kṣaumaḥ*, meaning hall or house on top of a rampart, from where enemies outside could be attacked: *prakārāgṛstha rānāgṛha* (Hari Charan Bandyopadhyaya-Dictionary), also meaning a watch-tower;

III. *pratoli*: gateway of a fort city (Bandyopadhyaya Dictionary, quoting *Amarkośa Ṭīkā*);

IV. *devapatha*: normally means heavenly passage, firmament (application of any one of these sense makes the entire discourse of the particular *sūtra* as meaningless. We have to follow here the suggested meaning of the word in relation to its placement on top of the rampart).

Adopting these senses of the words cited above, the *sūtras* mean to convey that watch-towers are to be constructed on square pillars (pillars having square cross-sections) at 30 *daṇḍa*s apart and are to be provided with a ladder-like staircase (a 'contrivance, generally portable, with rungs between two supports, for going up and down') for scaling up the height of the tower. Half way between the two towers, the entrance gate, with an apartment on top of it, is to be built. In between the space of the gate and the tower, notched planks for three archers (for operating for counterattack on enemies from outside) are to be provided as covering.

Two *hasta*s away from the pillar, roads twice this width have to be constructed along the top of the rampart.

Above translations of the *sūtra*s are comformatory to the concepts of different strategic constructions perceived for full protective measures for the fort.

7. ***Sūtra* 2.12.8**: Book 2, chapter 12 of *Arthaśāstra* deals with the description of ores and prescribes various processes and ingredients in treating such ores for extracting metals therefrom. It may be pointed out that purification, expressed by the word *śuddha* actually means extraction of the metal from its ore. That will dispel the confusion in the concept of purification and extraction. Obviously the removal of undesirable materials associated with the ore of the metal sought for gives the idea of purification, though this constitutes a part of the process of extraction.

The *sūtra* indicates the processes of digestion (*bhāvita*), preparation of pastes (*avalepa*), and calcination (*prativāpa*). The treatments have been prescribed for ores, either impure (*aśuddha*), or a low grade one (*mūḍhagarbha*). The ingredient to be used are ammonia (*tīkṣṇamūtrakṣāra*), bile urine and dung of animals, like cow, elephant, camel, buffalo, along with vegetative components derived from banyan tree, *rājavṛkṣa* (Pipal — *Buchanania Latifolia*, or milk hedge — *Euphorbia Tirucalli*), *pilu* (palm tree), etc.

Mūḍhagarbha in Sanskrit means 'a dead foetus' (M.M. Williams, p. 825). Use of this word in describing an ore appears to be rather confusing, though critical analysis would show that the use has been in the form of analogy, specially when read with the preceding *sūtra*, where it is stated that for all the ores, the higher density (heaviness) is indicative of higher quality (in terms of high concentration of metal content). This has been described drawing analogy from the biological process, that the gradual development of an embryo (*sattva* means also embryo, besides its other meaning of quality) is indicated by the heaviness of the body carrying it. In continuation, the state of an ore has been described using similar analogy. The word *mūḍhagarbha* indicating less developed, has been used in case of an ore, for conveying the meaning that it is immature in metal content, or low grade ore. Read in conjunction with *aśuddha* (impure), *mūḍhagarbha* employed to describe the state of an ore may be taken to mean a low grade ore. The process of extraction described in the *sūtra* must have been prescribed for low grade and impure ores.

Tīkṣṇamūtrakṣāra is a typical example of 'bound morphemes', comprising three free morphemes, *tīkṣṇa*, *mūtra* and *kṣāra*, the bound morpheme having a derived meaning of *kṣāra* (caustic) in liquid form having a strong acrid smell. In combination of the word *mūtra*, the word in all probability indicates ammonia, which was previously thought of as spirits of hartshorn.

Concrete nomenclature of a process or of a substance is a modern outcome of perfection of science. In the earlier stage of intellect, when empirical derivatives of cognition and observations were recorded, such descriptions mostly were pleonastic and tautological, devoid of conciseness and terseness. The latter, as stated also developed as a result of evolution of science and along with that the preciseness of the language of science. Analogies mostly resorted to in ancient literatures of natural science and philosophy, as in *Arthaśāstra*, has been to denote deeper-lying cases of resemblance, as in *sūtra*s 2.12.7 and 2.12.8. If the process thus described can be grasped properly, then there is hardly any difficulty in deriving meanings intended to be conveyed by such analogies. Kauṭilya himself indicated such a method of describing things or situations as *dṛṣṭena dṛṣṭasya sādhanamupamāṇam* (15.1.27). Such deconstructions may reveal wealth of information as in the case of deconstructing ancient scripts.

The *sūtra* 2.12.8 can then be taken to mean as: the impure and low-grade ores should be digested with ammonia, pellets formed thereafter with biles of cow, urine of buffalo, cow and camel mixed with vegetative ingredients of banyan, palm, *pipal* trees and milk-hedge, should be calcined with dung of these animals, in order to extract pure form of metals out of their ores

8. ***Sūtras* 2.12.15 and 2.12.16**: *khurumbaḥ pāṇḍurohitaḥ sinduvarapuṣpavarṇan va tīkṣṇadhātu* (2.12.15); *kākaṇḍabhuja-patravarṇau va vaikṛntakadhātuḥ* (2.12.16).

The *sūtra*s characterise two types of ores (metals) by their respective colours. The first one, the ore of *tīkṣṇa*, is of the 'colour of orange, or light reddish or of vermillion' (Jogindra Nath Samaddar, in Bengali, p. 95). The same *sūtra* has been translated by Kangle (Kangle, pt. II, p. 107) as 'that which is made up mostly of smooth stones, is whitish red or of the colour of *sindūvara* flower is iron ore'.

The second, *vaikṛntaka dhātu*, is the one which has the colour of *kanda* leaf or birch leaf (Samaddar). The same by Kangle (pt. II, p. 107) has been translated as 'that which is of the colour of *kākaṇḍa* (crow's egg) or birch leaf is *vaikṛntaka* ore'

Two consecutive *sūtra*s can be taken to be describing two types of ores of the same metal, viz., iron, though the exact identification of this metal is uncertain, according to Kangle (Kangle, pt. II, p. 108).

These two *sūtra*s evoked much imagination as will be evident from the following excerpts.

It is suggested 'that since magnetite is black and becomes brown on surface oxidation, being changed to hematite, *vaikrntaka* ore is magnetite' (Roy, 1946, p. 113). Kangle, on the other hand, quoting from a manuscript, supposed to be the compendium of two fragmentary Sanskrit commentaries, comes to the conclusion 'that the name is derived from some locality, and may be some special kind of iron or steel' (Kangle, pt. I, p. xvi, and pt. II, p.108). In another discourse it is said, 'We have already discussed the unsolved puzzle regarding the identity of *vaikrntaka*. All we know is that it was a metal or alloy prepared from its ore or *dhatu*, which according to Roy (1946, p. 113), could be magnetite. *Vaikrntaka* has been referred sometimes (2.12.23, 4.1.35) with *vrtta*, which is identified by many scholars, including Kangle, as steel. In one place (2.17.4) *vaikrntaka* and *vrtta* have been mentioned with *kalayasa*, which is evidently the Vedic name for pig or wrought iron; thus the three (2.17.14) were presumably believed to be different from each other, on the tope of it, *tikshna* mentioned as iron, had its ore or *dhātu* (2.12.15), and the metal was used as an alloying component (2.12.24, 2.13.48 and 2.13.58).

We propose that *kalayasa* stood for pig or wrought iron and *tikshna* was carburised iron (not cast iron) which gave sharp (*tikshna*) tools. *Tikshnadhatu* was possibly hematite. Iron prepared from south Indian magnetite or *vaikrntakadhatu* was wrongly believed to be different metal *vaikrntaka*' (Biswas and Biswas, 1996, vol. 2, pp. 51-2).

No doubt that both the *sūtras* deal with iron or its ore. *Tīkṣṇa*, from the very name indicates something very sharp and in all probability describes the product from the raw material, its ore. The latter, having the colour as described by Samaddar, which agrees more to the description of the *sūtra*, very closely resembles the major source of metallic iron in India, the Banded-Hematite-Jasper (BHJ), in the parlance of Indian Geology. This ore is composed of alternate bands of cherry-red coloured Jasper and steel grey coloured hematite. In the production of iron and steel, this ore is widely used for its availability in vast quantities with high iron content. The iron has not been named in the *sūtra*, but only the ore of it has been described (*dhātu* has also been described to mean *akāra*, ore). As regards *vaikṛntaka* several words are available in dictionaries, like *vikrānta* (a sham diamond, also bold, strong, bold advance), *vikrānti* (all-pervading power, strength), *vaikṛnta* (mercury) or *vaikranta* (a kind of gem resembling diamond). If we look into the etymology of the words, it may be seen that *vaikṛntaka* is derived from the root *vi+kṛnta+aca* (*ghan*), that which has been so tempered as to give rise to sharp product capable of cutting and shearing. *Vaikrantaka*, on the other hand, is derived from the root *vi+kṛm+aca* (*ghan*), conveying the sense of induction by external influence. Two interpretations are possible.

If the word *dhātu* at the end (2.12.16) is taken as metal, the word *vaikṛntaka*, in all possibility, means the processed metal, iron or steel. Otherwise, if *dhātu* is taken as indicating 'ore', this conforms to the physical description of magnetite, another ore of iron, having often properties of attracting magnetic materials like iron (loadstone). In any case, the word cannot be taken to mean mercury or diamond. *Vaikrantaka*, thus may indicate the loadstone.

9. ***Sūtra* 2.13.3**: This *sūtra* relates the occurrence of gold in two ways. The first part, the geographical localities like Jambū

river, Śatakumbha (Kangle refers this as a mountain, Meyer thinks this to be a river: Kangle, pt. II, p. 111), Hāṭaka, Mt. Venu and Śṛṅgasūkti have been indicated, from where gold could be obtained. We do not venture to identify the localities, as has been attempted by some authorities, with altogether different objectives, that is of locating these places with reference to the modern geographical set-up. We are contended only to accept that these localities only indicate the geographical occurrence of gold. The latter part of the *sūtra* actually indicates the genesis of gold — mainly in three forms, viz. (i) *jātarūpam* or in native form, (ii) *rasaviddha* and (iii) *akoradyatam*, or from ore. Of these, the second form has been described as 'that transmuted by liquids' (Kangle, pt. II, p. 111). This translation does not convey the real meaning, and a sense of transmutation, a process in alchemy, is silently conveyed, and is evident from the similar translations of the *sūtras* 2.12.2. and 2.12.3 (Kangle pt. II, p. 106). *Rasavidha* in the *sūtra* may mean the type of gold ore worked and reworked by water, as in the Placer deposit. Gold and other durable heavy minerals are derived from a sequence of events starting with (i) the preparation or release of such individual particles of metal or heavy minerals from enclosing rocks by prolonged weathering or by communition of such mineral bearing rocks; (ii) sorting and concentration of these by sorting agents, mainly flowing water, which carry them and deposit in favoured places. These favoured places are the stream channels, at places where stream velocity is slackened, along the flattened gradients of the stream or where the stream emerges into quieter water, along a broad valley below a narrow gorge (comparable to *bilaguhopatyakalayanagudakhatesvantaḥ* in *sūtra* 2.12.2), etc. There is indication of similar deposit in this *sūtra* itself, when *jambūnāda* is indicated as one of the sources. Thus, *rasaviddha* need not be considered as associated with mercury, as has been described elsewhere. There is no confusion about *akoradyatam* — since gold occurs in combination with other metals as ores, which require proper treatment for extracting the metal.

10. ***Sūtra* 2.24.5:** The *sūtra* reveals the rainfall zoning of the country, relevant to the agricultural practice, in which water is the predominant input for agricultural production and productivity. Two aspects are clearly indicated. One is the harvesting of rainfall, in time and space, and which varies from one region to another of the country, from only 13 *droṇa*s to unlimited. It is clearly suggested that sowing should be carried out according to the time and amount of available rainfall (*deśavāpa*, as opposed to *kulyavāpa*, which is artificial irrigation through canals; Kangle, pt. II, p. 149). The need to provide artificially water for sowing has been indicated as *haimanyanam* and *kulyavāpanam ca kalataḥ*, meaning providing water through canals during the winter (when there is no rainfall) and at other times when water is required for crops. This interpretation by Pran Nath (*The Indian Antiquary*, June, 1931, pp. 111) seems to stand to reasons. It explains well when the last part of the *sūtra* is segregated as a separate sentence. This can justifiably be done considering that this is a *sūtra*, which often is not a grammatically completed sentence, but can be so edited as to bring out the suggested sense and concept with reference to the context of the subject being dealt. The proposed construction, thus, stands as — *ṣoḍaśadroṇam . . . amitamparantanam: haimanyanam ca kulyavāpanam ca kalatah*.

Pran Nath's interpretation of this *sūtra* as indicating the share of crop that is due to the king, seems to be out of tune with the preceding and succeding *sūtras*, which deal with different aspects of agriculture, including rainfall zoning, rainfall harvesting, artificial canal irrigation. In continuity with the rainfall zoning, it is also stated that excellent production is ensured with the sum total of one-third of the annual rainfall in the first and last months and two-thirds during the intervening months (between sowing and maturing of corps) (2.24.6) It is further stated that it is possible to ascertain this sequence from the position of sun and constellations of other stars (2.24.7). All these have been made related to rainfall, as in *sūtra* 2.24.6. Sudden import of altogether a different sense to a *sūtra* interposed in between other *sūtra*s dealing with a particular subject and having conceptual continuity appears to be a literary anamorphosis.

11. ***Sūtras 3.8.2 to 3.8.5***: These *sūtra*s describe the components of a unit dwelling place for individual citizen, and the mode of construction of a house. *Sūtra* 3.8.2 depicts a unit dwelling place (*vastu*) as comprising of house (*gṛha*), adjacent land (*kṣetra*), a garden (*aramaḥ*), and embanked pond (*setubandhastatakamadharo*). In *sūtra* 3.8.3, the word *setu* has been defined. It is the process of tying the beams with the roof by iron hook (*karṇakīlayasa* — bolt bent at one or either ends, resembling ear). Meyer also means by it as 'an iron joining with pegs (*kila*) that have ears (*karṇa*, i.e., barbs)' (Kangle, pt. II, footnote, p. 216). Since *setuḥ* has thus been defined in original *sūtra*, its use in the sense conveyed in *sūtra* 3.9.10 need not be invoked, as preferred by Kangle. Accordingly, the house is to be built depending on the enduring capacity (*bhoga*) of the beams, and the foundation (*pade bandham*) for the house has to be two *aratnī* by three *padas*, and away from the boundary of other house (*parakuḍya*), even if that is not demarcated (*abhūtam*) by a wall. The didactic sense of these *sūtra*s appears to have been brought out in more clearly by the translation of Samaddar in Bangali (Samaddar, BS 1319, pp. 185-6), which reads as follows:

'The process of trying beams with roof by iron bolts is known as *setuh*. The house is to be built on ascertaining the durability of the *setuh*. New house is to be built without encroaching upon other's land. The foundation will have to be two *aratni* long and three *pada*s wide.'

It would be of interest to note the mode of construction of buildings of that time as described by Ghosh (1919, p. 25):

> 'The superstructure of the building was most likely formed of heavy logs of *sal* wood which appears to have rested directly on the columns without the intervention of stone capitals, and to have been held in position by heavy round copper bars. The timbers were themselves fastened together by extraordinarily large massive nails, six to eight inches or more in length.'

Another aspect is necessary to be taken into cognition in support of the above deconstruction of the *sūtras*, and for disconsidering the one by Kangle. That relates to the antiquity of the technology of manufacture of iron nails and wires.

Kangle lays more emphasis on the word *setuḥ* as meaning 'boundary or boundary mark', demarcating the 'expanse, extent' (*bhoga*) of one's dwelling place, and translates the *sūtra* 3.8.3 as, 'along the house (is to be) a boundary, fixed with iron wires (*sic*) in pillars at the corners'. The *sūtra* does not contain any such words, which could be translated as 'iron wire' and 'pillars'. These words have been introduced in the translation only to establsih the concept of boundary around dwelling places.

If little attention is paid to the history of technology, we would notice that though the wires, made from such ductile metals as gold, silver or copper were known to be in use in ornaments and filigrees, no such use was known of iron wires. Iron wire was made by forging until the tenth century, when the draw-plate was invented' (Charles Singer and Holmyard, 1979, p. 75). On the contrary 'both the nail and rivet were known to the coppersmiths of Egypt well before 2500 BC. The Iron Age naturally brought with it, among other things, iron nails, but for large timbers, such as are used in ships, where these were to be joined, it was common to use trenails. . .' (Ian McNeil, 1990, p. 14). Spiral headed pins have also been discovered from Mahenjo-Dāro, Harappā and Chanhu-Dāro, which consisted of a copper or bronz pin in which an ornamental head was contrived either by splitting the wire at the top and twisting it into two flat spirals side by side, or in which a single spiral was made by turning over the top of the pin without splitting the metal (Piggot, 1947-8, pp. 26-7). Similar nails with hook head have been found from Taxila, belonging to first century BC to second century AD (Ghosh, 1944-5, pp. 41-84).

Thus, while use of nails, bolts and rivets was known from much earlier date, the type and size of wires, as are necessary to put boundaries round a dwelling place could not have been manufactured in such dimensions and abundance as could be available for use by ordinary citizens. On this account also, the translation of *sūtra* 3.8.3, as offered by Kangle does not reveal reality, but dwells on speculative overtones. On the contrary, Meyer (as quoted by Kangle, pt. II, footnote, p. 216) hinted to it, and Samaddar used it to convey the most plausible translation, which fits well with the context and intended objectives, as related to house building.

Glossary

(The figures within parenthesis indicate successively the Book, chapter and *sūtra* numbers, as in Kangle's translation.)

Aharyodakam (2.1.20): Water that is obtained artificially from the reservoirs created by dams, or through channels from rivers or other sources.

Aratnī (2.5.7): This was an ancient measure of length from the human elbow to the tip of his little finger, normally taken as equivalent of 18 inches. In this *sūtra*, the size of the mouth of a pitcher to be used for measuring the amount of rainfall has been indicated to be one *aratnī*, or 18 inches. If this measure is taken to indicate the circumference of the pitcher (*kuṇḍa*), it compares well with the standard rain-gauge of present day, where the funnel receiving the rainfall has diameter of 127 mm, that is circumference of about 399 mm, which is equivalent to about 16 inches.

Arcismanantargata pravāḥ (2.11.33): Refracted light from within a gem. In describing the characters of gems, it is stated that the one with intense internal luminescence/refraction, belongs to the excellent category of gems.

Arthaśāstra: This is a compound Sanskrit word of *artha* and *śāstra*. The word *śāstra* is variously used to denote any branch of knowledge, a scientific work, or a theological treatise. *Artha* literally means wealth, the present cannotation of *Arthaśāstra* is science of economics. But in ancient Sanskrit literatures, *Arthaśāstra*, *Daṇḍanīti* and *Rājaśāstra* or *Rājnīti* are synonymously used. All these come within the purview of Dharmaśāstra in Sanskrit literature, which covers a wide range of meaning and knowledge. The word *dharma* having various meanings like ordinance, usage, duty, right, justice, morality, virtue, religion, good work, function, characteristic, etc., passed through several transitions of meanings, and the significant meaning established were principles, duties and obligation of man, his standard of conduct as a member of Āryan community, as a person in a particular stage of life. Dharmaśāstras deal with all aspects of society and hence with Rājadharma also (*rājā* meaning the king). *Daṇḍa* (cf. *Daṇḍanīti*) literally means a stick or staff for driving cattles, but came to mean 'the power of control of punishment', that is, it conveys the power of the king or supreme ruler, called *Daṇḍadhara*, to dispense justice and to punish wrong doers, or in other words, the science of government. Kauṭilya refers it as *Asthaśāstra*. *Artha* is material well-being, wealth, played in man's life. Wealth depended on land and labour. Unless there existed some authority that regulated the people and their doings, there would have been chaos and no security. Hence, the ruler or the king was the most important institution in the society, and the rules for the king's guidance were called *Rājadharma* or *Rājanīti*. *Arthaśāstra*, therefore, came to be identified with *Rājanītiśāstra*, or simply *Nītiśāstra* (Kane, P.V., *History of Dharmaśāstra*). Kauṭilya states that by combining the essence of all such scriptures written by his predecessors on the art of aquiring the landed wealth and its maintenance (*pṛthivyā lābhe pālane ca . . . sūtra* 1.1.1), he has composed this one. His philosophy on the art of maintenance by the possessor (the king) is expresed as *prajā sukhe sukham rājñaḥ prajānām ca hite hitam, nātmā priyam hitam rājñaḥ prajānam tu priyam hitam* (1.19.34). This means that king's prosperity depends on the well-being and contentment of his subjects. The king does not gain from self-gratification or desire, but by satisfying the desires of his subjects. A nascent concept of welfare state even in those days of monarchical system!

Aṭṭālaka (2.3.10-2.3.13): Normally means a palace, an apartment on the roof, an upper storey. From the description of the structure associated with ramparts around a fort, having removable artefact for climbing up and coming down (may be a ladder-like device), *aṭṭālaka* means here a watch-tower with chamber on top of it.

Ātyantavarṇagauravam (2.12.1): Loud, bright colour, as imparted to soil/rock by chemical leaching of some mineral compounds, e.g., brilliant green colour by leaching of copper carbonate mineral, brown or reddish brown colour by hydrous iron oxide, etc.

Avarūdham (2.3.5): Firmly enclosed. The word has been used in connection with the construction of earthen rampart around a fort to indicate that the foundation below the ground is to be firmly enclosed to render it the necessary strength to support the superstructure and ensure its stability.

Ayam (2.3.18): Has been used to indicate both length and breadth.

Bamataḥ pradakṣiṇa sopānam (2.3.22): Sinistrally spiralled staircase.

Caraka pāṅśu dhabaka (2.13.33 and 4.1.49): *Dhabaka* normally means a washerman. *Caraka*, derived from *car*, meaning, to roam about, wander, and also, engaged in, busy with. It conveys the latter meanings in this *sūtru*. *Pāṅśu* means sand, in this case sand mixed with gold particles. Thus, *caraka pāṅśu dhabaka* means one engaged, in washing sand for recovering gold particles. The share of gold to be had having been indicated, the recipient is naturally the one who recovers it, that is the *dhabaka*.

Daṇḍa (2.3.5): A linear measure, being about 6 feet.

Dhātu (in many *sūtras* in Book 2, chapter 12): Variously been used to indicate both metal and ore.

Droṇa (2.24.5): Measure of volume of liquid, and is equivalent to 1/20th part of a *kumbha* (pitcher) (*sūtra* 2.19.32). Taking the capacity of a *kumbha* to be about 3 gallons (*Sanskrit-English Dictionary* by Monier-Monier Williams), one *droṇa* works out to be about 43 cubic inches (704644 cubic mm).

Gauravabr̥dhau sattvabr̥dhi (2.12.7): While describing the quality of ores, it is stated that increase in heaviness indicates better quality of ore. That higher proportion of metallic content of an ore is indicated by the higher specific gravity is self-evident.

Gūḍakhātaḥ (2.12.2): Deep gorge, cavern or crack.

Hasta (2.3.13): Measure of length from the human elbow to the tip of his middle finger, about 18 to 22 inches.

Indrakīlaḥ (2.3.26): An iron bolt.

Indranīla: Sapphire.

Jantrabādhā talabapatam (1.20.2): Platform, within a room, fitted with mechanical device to lower and raise it (comparable to a modern lift in a multi-storeyed building).

Jātarūpam (2.13.3): In the native state. Of various states of occurrence in nature, gold is found also in native state.

Kalka: Levigated powder.

Karṣanayantra (2.24.3): A plough.

Kṣāra (2.12.8): Caustic alkali.

Kuḍya (3.8.5): A wall. From this is derived *parakuḍya* meaning a wall at the farther end, or a boundary wall.

Mohana-gr̥ha (1.20.2): A love chamber.

Padmarāga (2.11.29): The gem Ruby.

Paṅka-mala (2.12.3): *Paṅka* means mud, mire, clay, etc., and *mala* means dirt, filth, impurity. *Paṅka-mala* thus means impure substances.

Pratoli (2.3.11): A gateway to a fort/city.

Puṭi kittaḥ (2.14.23): *Kittaḥ* means sediment, dirt, secretion, and *puṭi* means putrid. The combined word in this *sūtra* means the left over, discarded material in the metallurgical treatment of a metallic ore in a crucible, that is slag, in metallurgical parlance.

Raṣaḥ kañcanika (2.12.2): In some English translations, the combined word has been used to indicate 'gold bearing liquid'. But in the succeeding *sūtra*, the character of this liquid has been described as 'when thrown in water spreads on the surface like oil, and absorbs mud and dirt'. From the processes described it appears to be like an electrolyte acting to coagulate the solid phase of a colloidal solution. Taking this cue, it has been interpreted in the present work that *raṣaḥ kañcanika* represents the juice of a plant, locally called *kañcanika*, which is related to the genus Premna, and which might be effective as an electrolyte for a colloidal solution containing fine gold particles with sand and clay, as in a sediment laden stream water.

Śālajatu (2.12.4): A mineral product found in Nepal and called in local dialect as 'Passeo', the sweat of rock. It is commonly believed that the material is exuded from rocks. Locally these are in great demand for medicinal purpose. On analysis the material was found to be a sulphate of alumina, tastes saline and is soluble in water.

Setu bandha (2.1.20, 7.12.4): *Setu* means a ridge of earth, mound, bank, embankment, any piece of land separating adjacent fields (cultivated), etc., *Setu bandh* means the process of constructing a dam across a river.

Tīkṣṇamūtrakṣāra (2.12.8): Ammonia.

Tutha (2.13.10): Has different meanings, like fire, stone, crucible. The latter meanings has been adopted in the same chapter, as *asthitutha* and *śuskatutha* meaning crucible made of earth mixed with bone powder and earth mixed with sand or earth alone respectively.

Ugragandharasam (2.12.1): Acrid, pungent smell of aqueous solution, derived from decomposition of some mineral and sulphide ores.

Vaidūrya: Beryl.

Vaidyutena bhasmana mṛt (1.20.4): Ash produced by lightening. The descriptive usage in consonance with the process described in the *sūtra* indicates it to be pumice (brick) produced by excessive fire-heating.

Vāpa (2.24.2): The act of sowing.

Viṣkambha (2.3.5): Has various meanings like post, pillar, prop, supporting beam of a house, etc. This word has been used in almost all these senses in different *sūtras*.

Bibliography

Acharya, P.P., 1927, *A Dictionary of Hindu Architecture*, Oxford University Press.

Adams, F.D., 1938, *The Birth and Development of Geological Science*, Dover Publication Inc.

Agrawala, V.S., 1963, *India as known to Panini*, Prithvi Prakashan, Varanasi.

Alexeev, V.P., 1986, *The origin of the human race*, Progress Publishers, Moscow.

Alice Jacob and S.N. Singh, 1972, *Law relating to irrigation*, Indian Law Institute, New Delhi, N.M. Tripathi Pvt. Ltd., Bombay.

Allchin Bridget and Allchin Raymond, 1968, *The birth of Indian civilization, India and Pakistan before 500 BC*, Penguin Books.

Allchin, F.R., 1962, 'Gold mining in ancient India', *Journal of the Economic and Social History of the Orient*, vol. 5.

Allchin, F.R., 1995, *The archaeology of early historic South Asia — emergence of cities and states*, Cambridge University Press.

Altekar, A.S. and V.K. Mishra, 1959, *Report on Kumrahar excavations (1951-53)*, K.P. Jaiswal Research Institute, Patna.

Andre Mayer and Jean Mayer, 1974, 'Agriculture, an island empire', *Daedalus*, vol. 103, no. 3.

Apte, V.S., 1989, *The Practical Sanskrit-English Dictionary*, Motilal Banarsidass.

Bakshtanovsky, V.I., *et al.*, 1989, *Ethics*, Students Library, Progress Publishers, Moscow.

Ball, V., 1881, *A manual of the Geology of India-Economic Geology*, Geological Survey of India.

Basham, A.L., 1954, *The wonder that was India*, Grove Press Inc.

Bernal, J.D., 1944, *The social function of science*, George Routledge and Sons Ltd. (4th reprint).

Bernal, J.D., 1969, *Science in history*, vols. 1-4, Penguine Books.

Bhardwaj, H.C., 1979, *Aspects of ancient Indian technology*, Motilal

Banarsidass.

Bhat, M. Ramkrishna, 1997, *Varahamihir's Bṛhat Saṁhitā*, Motilal Banarsidass.

Bhattacharya, Saktiranjan, *et al.*, 1994, 'Vastu Vidya in ancient India', in *Science and Technology in ancient India*, ed. Manabendu Banerjee, Sanskrit Pustak Bhandar, Calcutta.

Bhattacharya, T.P., 1986, *The Canons of Indian art or a Study of Vastuvidya*, Firma KLM Pvt. Ltd., Calcutta.

Biswas, A.K. and Sulekha Biswas, 1996, *Minerals and Metals in Ancient India*, vols 1-2, D.K. Printworld (P) Ltd., New Delhi.

Bronowski, J., 1960, *The Comrnon Sense of Science*, Pelican Books.

Brooks, B.R., 1972, *Geobotany and Bio-geochemistry in Mineral Exploration*, Harper and Row Publishers.

Campbell, A., (i) 1833, 'On the Native Alum or Salajit of Nepal', *Journal of The Asiatic Society*, part II. (ii) (1834), 'On the Nepalese method of refining gold', *Journal of The Asiatic Society*, part III.

Charles, Singer and E.T. Holmyard, 1979, *A history of Technology*, Oxford University Press.

Chattopadhyaya, B.D., 1984, *A survey of historical geography of India*, Manisha, Calcutta.

Chattopadhyaya, D.P. , 1986, *History of science and technology in ancient India — the beginnings*, Firma KLM Pvt. Ltd., Calcutta.

Chattopadhyaya, P.K., 1992, Archaeo-metallurgy in India: studies in technoculture in early copper and iron ages in Bihar and West Bengal: unpublished Ph.D. thesis, Patna University.

Chatterjee, S.K., 1974, *Kirata-Jana-Kriti*, The Asiatic Society, Calcutta.

Chaturvedi, R.C. and M.M. Chaturvedi, 1996, *Law on protection of environment and prevention of pollution*: The Law Book Company (P) Ltd., Allahabad, (2nd print).

Childe, V.G., 1956, *Man makes himself*, Watts & Co., London (3rd edn.).

Das Gupta, S.P., 1995, *Environmental laws in ancient India*, Annual Volume, Centre for the Study of Man and Environment, Calcutta.

Das Gupta, S.N., 1987, *Natural Science of ancient Hindus*, ed. Debi Prasad Chattopadhyaya, Motilal Banarsidass.

Datta, Nalinaksha, 1953, 'Buddhism' in '*The History and Culture of the Indian people — the age of imperial unity*,' Bharatiya Vidya Bhavan.

Department of Scientific and Industrial Research, Road Research Laboratory, U.K., 1959, *Soil mechanics for road engineers*, Her

Majesty's Stationary Office, U.K.

De, Sudhin and P.K. Chattopadhyaya, 1989, 'Iron objects from Pandurajar Dhibi-archaeometallurgical studies', *Steel India*, vol .12, no. 1.

Dey, R.C., 1983, 'A note on the ancient smelting in South-western Haryana: Indian Minerals', *Geological Survey of India*, vol. 37, no. 3.

Donald, E. Strong, 1965, *Landmarks of the world's art — The classical World*, Paul Hamlyn, London.

Dudich, Endre, 1984, *From alchemy through geochemistry to cosmo-chemistry*: Proceedings of the 27th International Geological Congress, vol. 21: VNU Science Press, Utrecht, The Netherlands.

Dwivedi, D.N., 1964, *Geographical data in Arthasastra*; XXVI International Congress of Orientalists, New Delhi, Bhandarkar Oriental Research Institute, Poona.

Encyclopaedia Britannica, Micropaedia, 15th edn., 1985.

Eric, Weil, 1975, 'What is breakthrough in history?', *Daedalus*, vol. 104, no.2.

Erich, W. Zimermann, 1951, *World resources and industries*, Harper and Brothers.

Foster, John Bellamy, 1995, *The vulnerable planet — a short history of the environment*, rpt. Cornerstone Publications, Kharagpur, West Bengal, from original publication by Monthly Review Press, New York.

Filliozat, Jean, 1969, 'Technology in India', in *A history of technology and invention: Progress through the ages'*, ed. Maurice Daumas, vol. 1, Crown Publishers Inc., New York.

Fustel, de Coulanges, 'Introduction to the history of the political institutions of ancient France', quoted by judith N. Shklar in 'Learning without knowing', *Daedalus*, vol.109, no.2, 1980.

Geoge Perkins Marsh, 1864, Man and Nature.

Ghosh, A., 1944, Taxila (Sirkap), *Ancient India*, no.4, Archaeological Survey of India.

Ghosh Manoranjan, 1919, Pataliputra, Patna Law Press.

Ginsburg, I.I., 1960, *Principles of geochemical prospecting*, Pergaman Press.

Gupta, S.P., 1980, *The roots of Indian art*, B.R.Publishing Corporation, Delhi.

Haldar, Aruna and S. Sen Sarma, 1994, Kautilya's Arthasastra in the light of modern science of minerals and mining, *Vishveshvaranand Indological Journal*, vol. XXVII, parts i-ii, June-December, 1989.

Hall, Rupert and Marie Hall, Boas, 1964, *A brief history of science*, Signet Science Library.

Havell, E.B., 1915, *The ancient and medieval architecture of India*, John Murray, London.

Hegde, K.T.M., 1991, *An introduction to ancient Indian metallurgy*, Geological Society of India.

Indian Minerals, vols. 13 & 16 (1959 and 1962), Geological Survey of India.

Jacobi, Hermann, 1918, *On the authenticity of Kautiliya': Indian Antiquity*, vol. 47.

Jain, Ramchandra, 1972, *McCrindle's Ancient India as described by Megasthenese and Arrian*, Today and Tomorrow Printers and Publishers, New Delhi.

Jevons Stanley, W., 1900, *The Principles of Science*, Macmillan & Co. Ltd., London.

Jolly, J, 1923, *Arthasastra of Kautilya*, vol. 1, Motilal Banarsidass.

Kangle, R.P., 1986, *The Kautiliya Arthasastra*, Parts 1-3 (rpt. 1986), Motilal Banarsidass.

Kane, P.V., 1968, *History of Dharmasastra*, vol.1, Bhandarkar Oriental Research Institute, Poona.

Karl Marx, Capital.

Klaus, Michael, Abich Meyer, 1996, 'Humans in nature — towards a physiocentric philosophy', *Daedalus*, vol. 125, no.3.

Kurien, N.P., 1992, *Design of foundation system — principles and practices*, Narosa Publishing House, New Delhi.

Lewis, Mumford, 1944, *The culture of cities*, Seeker and Warburg, London.

Louis, Dumont, 1975, 'On the comparative understanding of non-modern civilization', *Daedalus*, vol.104, no.2.

McKinstry, H.E., 1955, *Mining Geology*, Prentice Hall, Inc.

Mao, Tse Tung, 1963, *Where do correct ideas come from? Selected Writings*, National Book Agency, Calcutta.

McNeil, Ian, 1990, *An encyclopaedia of the history of technology*, Routledge.

Mark, Svendsen and Small Less, 1990, *Assessing irrigation system performance*, Overseas Development Unit of Hydraulic Research Bulletin, November 18, 1990; Wallingford, Oxfordshire.

Milovsky, A.V. and Konov, O.V., 1985, *Mineralogy*, Mir Publishers, Moscow.

Mind Alive, 1970, *Marshal Cavendish Encyclopacdia*, Hamlyn Publishing Group.

Mitra, A.K., 1927, 'Mauryan art', *Indian Historical Quarterly*, vol. III, 1st rpt., 1985, Caxton Publications, New Delhi.

Monier Monier Williams, 1963, *Sanskrit-English Dictionary*, Oriental Publishers, Delhi.

Mookherjee, Asoke, 1992, *Current Science*, vol. 63, no.4.

Mukherjee, R.K., 1953, 'Magadhan Imperialism' in *The history and culture of the Indian people*, vol. II; Bharatiya Vidya Bhavan.

Nag, K., 1923, Quoted by Kangle, R.P., in *The Kautiliya Arthasastra*, II, University of Bombay, 1965, pp. 74.

National Council of Educational Research and Training, India (1985), Agricultural meteorology.

Pascoe, E.H., 1965, *A manual of the geology of India and Burmah*, 3rd edn., Government of India.

Peter Rose, Myron Glazer and Perina Glazer, 1982, *Sociology, inquiring into society*, St. Martin's Press, New York (2nd edn.).

Piggot, Stuart, 1948, 'Notes on certain Pins and Mace-head in the Harappa culture', *Ancient India*, no.4, Archaeological Survey of India.

Prasad, A.K. and P.K. Chattapadhyaya, 1987-8, 'Protohistoric copper objects from Taradih', *The Journal of The Bihar Puravid Parisad*, vols. XI & XII.

Pran Nath, 1931, *Indian Antiquary*, June, 1931.

Rangarajan, L.N., 1992, *Kautiliya — the Arthasastra*, Penguin Classics.

Robert, L. Bases, and Julia Jackson, 1987, *Glossary of Geology*, American Geologica Institute.

Roy, H.C., 1925, 'The date of Kautiliya', *Indian Antiquity*, vol. 47.

Roy, P., 1956, *History of chemistry in ancient and medieval India*, Indian Chemical Society, Calcutta.

Roy, P.R., 1986, *Cultural heritage of India*, vol.VI; *Science and Technology*, ed. P.R. Roy and S.N. Sen: Ramakrishna Mission Institute of Culture, Calcutta

Roy, S.K., 1946, 'Glimpses into India's Heritage in Geology, Mining and metallurgy', *Quaterly Journal of Geology, Mining and Metallurgy*, vol.18.

Roy Choudhury, H.C., 1958, *Studies in Indian antiquities*, Calcutta University.

Samaddar, Jogindranath, 1319, *Arthasastra*, Part 1 (in Bengali), D.N. Lahiri, Howrah, West Bengal.

Sarkar, Benoy Kumar, 1928, 'The German translation of Kautiliya Arthasastra', *The Indian Historical Quarterly*, vol. IV, 1st rpt., 1985, Caxton Publications, New Delhi.

Sethna, K.D., 1989, *Ancient India in a new light*, Aditya Prakashan, New Delhi.

Sen, Amartya, 1997, 'Indian traditions and the western imagination', *Daedalus*, vol. 126, no. 3.

Sen Sarma, Priyadarsan, 1998, *Ethnobiological information in Kautiliya Arthasastra*, Naya Prakash, Calcutta.

Sen Sarma, S.B., 1979, *'Engineering Geology — history of development, present and future status, Indian Journal of Power and River Valley Development*, May, 1979.

Sen Sarma, Sunil, 1998, 'Contemporaneity of perception on environment in Kautiliya's Arthasastra', *Indian Journal of History of Science*, vol. 33, no 1.

Sharma, Sukhomoy, 1983, *Society in Mahabharata* (in Bengali), Visva Bharati.

Sharma, Y.D., 1950, 'Building laws in Arthasastra', *Journal of the Oriental Research*, vol. XXI, 950- 1.

Spooner, D.B., 1913, *Archaeological Survey of India — Annual Report*, 1912.

Terzaghi, K. and Peck, R.B., 1967, *Soil mechanics in engineering practice* (2nd edn.), Wiley International Edition.

Thapar, Romila, 1975, 'Ethical, Religion and Social protest in the First Millennium B.C. in Northern India', *Daedalus*, vol. 104, no.2.

Tomlinson, M.J., 1963, *Foundation design and construction*, Sir Isaac Pitman and Sons Ltd.

Toulmin, Stephen, 1977, 'From form to function', *Daedalus*, vol. 106, no.3.

Van Buitenen, J.A.B., 1975, *The Mahabharata*, The University of Chicago Press.

Van Pelt, Robert Jan, and Carrol William Westfall, 1991, *Architectural principles in the age of historicism*, Yale University Press.

White, Robert M., 1996, 'Preface', *Daedalus*, vol. 125, no.3.

Winternitz, M., 1967, *History of Indian literature*, vol III, pt II; tr. Subhadra Jha, Motilal Banarsidass.

World Bank Policy Paper, 1993, *Water resources and industries*, Harper and Brothers Publishers, New York.

General Index

Index of Important Sanskrit Words